Customizing SolidWorks For Greater P

Written by: Joe Bucalo and Neil Bucalo

Published by
Sheet Metal Guy, LLC
P O Box 498283
Cincinnati, OH 45249
www.SheetMetalGuy.com

Please visit our website at www.SheetMetalGuy.com. As the reader of this book, you are our most important critic. You can email us at books@SheetMetalGuy.com to let us know what you did or didn't like about this book. We will carefully review your comments and share them with those whom helped make this book possible.

Manufactured in the United States of America.

All technical illustrations and CAD models in this book were produced using SolidWorks 2007. However most of the information applies to previous versions as well.

About the Authors

Joe Bucalo is the founder and President of Applied Production, Inc. and Sheet Metal Guy, LLC, SolidWorks Solution Partners. He has over 30 years experience working with and developing software products in the CAD/CAM world. When CAD was just beginning on the PC, Joe played a major role in the development of ProFold, the first truly automatic 3D sheet metal unfolding program. Joe later paved the way for graphics based sheet metal CAM when he introduced ProFab, which allows the direct transfer of geometry from CAD to CAM. He has a thorough knowledge of the most popular CAD programs and understands the issues faced by manufacturers.

Neil Bucalo is a CAD expert, having many years of experience using numerous CAD systems, including SolidWorks. Neil has a diverse background, including mechanical engineering, CAD/CAM support and training, and technical writing. He started his career in support and training at Computer Aided Technology, Inc., becoming a Certified SolidWorks Support Technician. Neil now works with his Uncle Joe, providing customer support and writing technical training books. Dedication: I would like to thank my lovely wife Becky for her loving support and patience. I am honored to have the opportunity to spend my life with my best friend.

Introduction

Customizing SolidWorks for Greater Productivity uncovers the power of SolidWorks by explaining in detail various customizing techniques available in the program. The purpose of this book is to provide you with information on customizing features like menus, toolbars, keyboard shortcuts, macros, templates, and views. It is my intention to get you thinking about how you can apply these features and other related features to enhance your personal experience with SolidWorks. Each chapter is written in a cookbook style, guiding you step-by-step to teach you as many techniques as possible in the least amount of time. Coverage is aimed at showing you how to become more productive by learning how to minimize repetitive actions, running SolidWorks more efficiently.

It is assumed that you already have a working knowledge of SolidWorks and the menu structure. You may want to open SolidWorks and in the Help menu, go through the Online Tutorial. Dialog boxes, toolbars, and buttons are shown throughout the book. When several buttons appear in a dialog box, the one which you should select is circled in the picture in the book. A circle will not appear on your SolidWorks screen. Before you start, take some time to familiarize yourself with the various parts of the SolidWorks interface. You will be instructed to access commands from different parts of the interface throughout this book. Refer back to this interface image if you get lost.

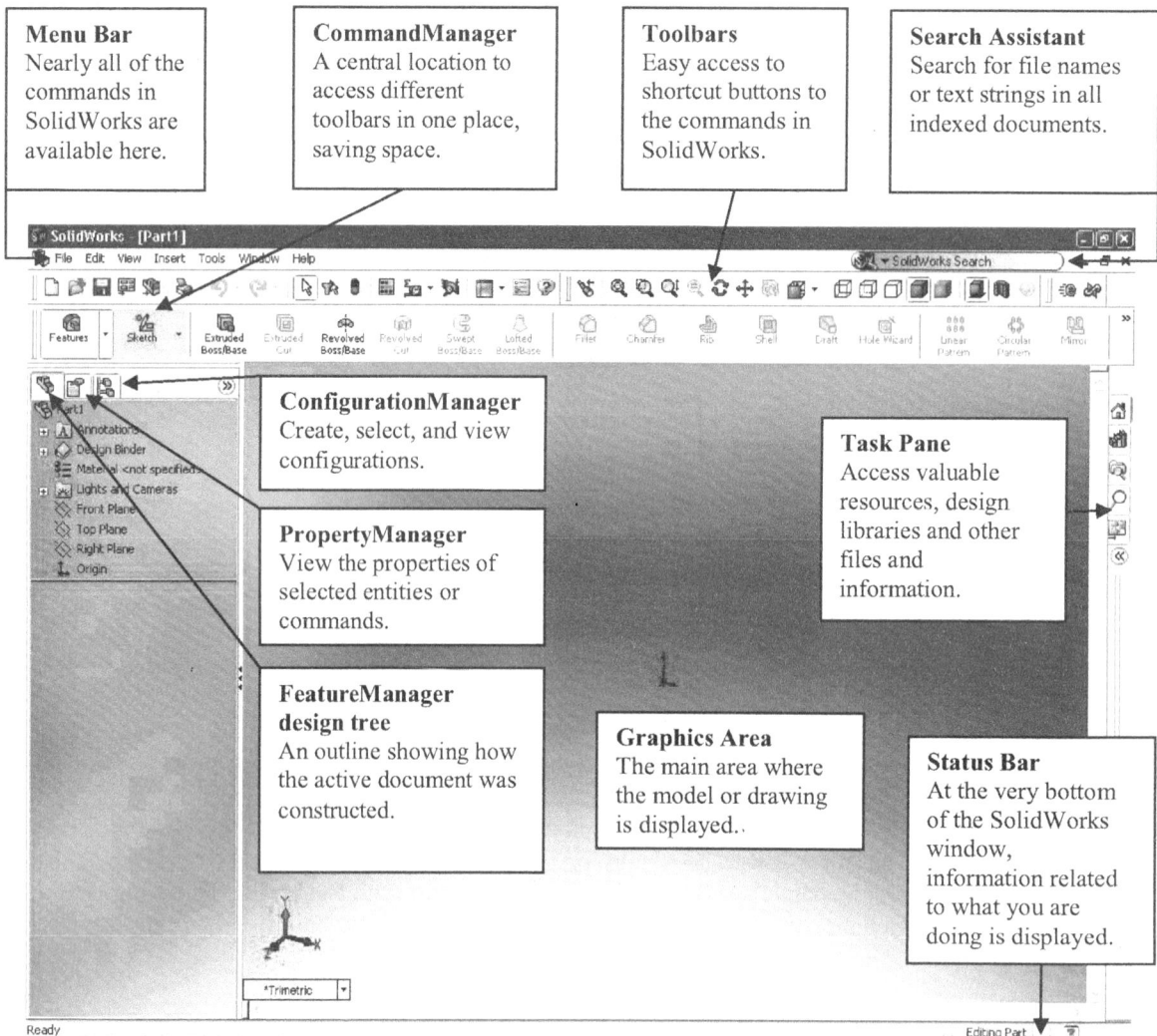

Menu Bar
Nearly all of the commands in SolidWorks are available here.

CommandManager
A central location to access different toolbars in one place, saving space.

Toolbars
Easy access to shortcut buttons to the commands in SolidWorks.

Search Assistant
Search for file names or text strings in all indexed documents.

ConfigurationManager
Create, select, and view configurations.

PropertyManager
View the properties of selected entities or commands.

FeatureManager design tree
An outline showing how the active document was constructed.

Task Pane
Access valuable resources, design libraries and other files and information.

Graphics Area
The main area where the model or drawing is displayed.

Status Bar
At the very bottom of the SolidWorks window, information related to what you are doing is displayed.

Table of Contents

Chapter 1

Save and Restore Your Settings

Since this is a book about customizing SolidWorks, you will make a lot of changes to the settings and display of your SolidWorks. To get started, you will need to backup your current settings so that no matter what changes you make, you can go back at any time and change everything back to what you had when you started. Since your current settings will be saved, you can play around without any worries of breaking anything. SolidWorks remembers all of the changes that you make as you try different system options and display changes. If things get messed up or you just want to get back to where you started, you can restore your setting with a few clicks.

You can use the Copy Settings Wizard to save and restore the system options, keyboard shortcuts, menu customization, and toolbar layouts. If at any time while working through this book, you need to reset your system and do some real work, just exit SolidWorks and use the Copy Settings Wizard to save your settings and restore your original settings.

Save Your Settings

Make sure that SolidWorks is closed. If SolidWorks is open, pull down the "File" menu and pick **Exit**.

Click on the Windows Start button and pick **All Programs – SolidWorks 2007 – SolidWorks Tools – Copy Settings Wizard**.

In the **SolidWorks Copy Settings Wizard** dialog box, click the **Save Settings** button, and then click **Next**.

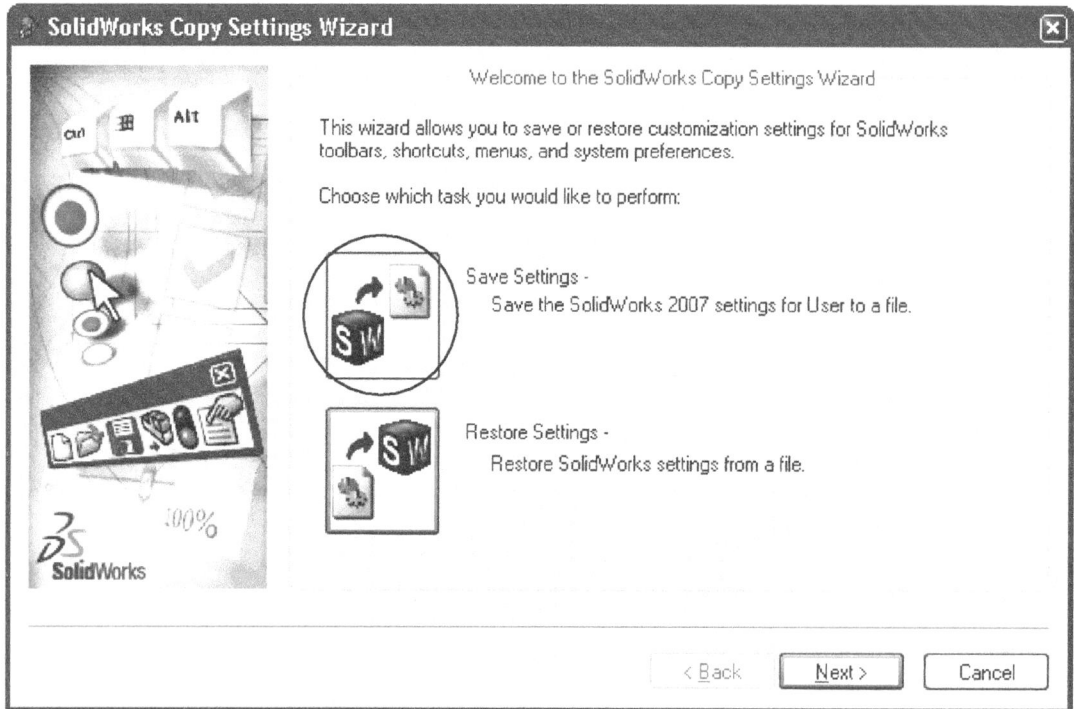

In the **SolidWorks Copy Settings Wizard** dialog box, use the default name and location for the settings file. Make sure that you take note of where this file is located and the name of the file so that in the future you will remember that this is the file to restore your settings back to the original setting that you had. The settings files have a default extension of **.sldreg**.

Under **Select which settings to save to file**, make sure that all four check boxes are checked and the **All toolbars** radio button is selected, and then click **Finish**.

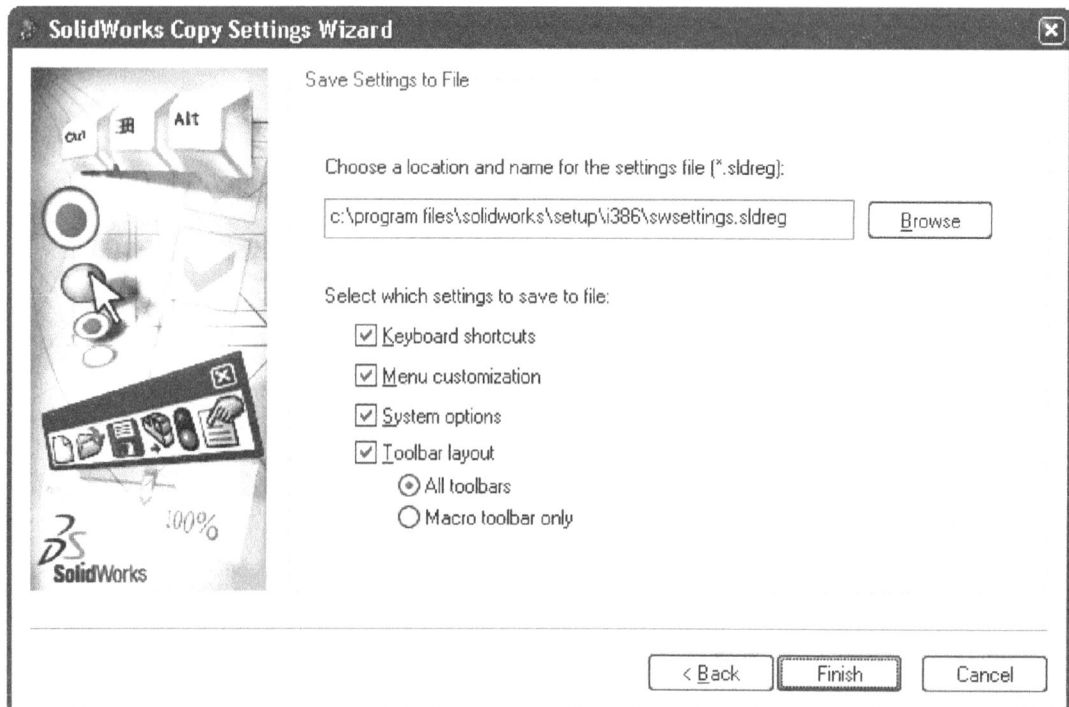

A dialog box will appear confirming that the settings have been successfully written to the **swsettings.sldreg** file and saved to the chosen location. Click **OK**.

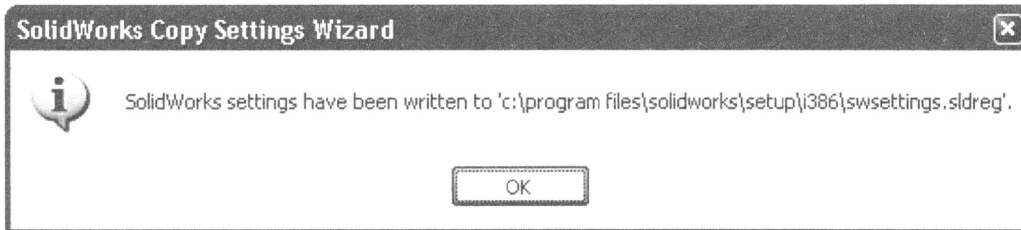

Restore Your Settings

Make sure that SolidWorks is closed, and then click on the **Start** button and pick **All Programs – SolidWorks 2007 – SolidWorks Tools – Copy Settings Wizard**.

In the **SolidWorks Copy Settings Wizard** dialog box, click the **Restore Settings** button, and then click **Next**.

In the **SolidWorks Copy Settings Wizard** dialog box, browse to the settings file (swsettings.sldreg) from which to restore the settings and pick the file. I told you during the **Save Settings** operation that you need to remember the file name and location to restore it.

Under **Select which settings to restore from the file**, make sure that all four check boxes are checked, and then click **Next**.

Chapter 1: Save and Restore Your Settings   3

In the **SolidWorks Copy Settings Wizard** dialog box, under **Select the Destination**, click on the **Current user** button, and then click **Next**.

Select the Destination

Current user (User)
Recommended for end users. Copies settings to the
CURRENT USER registry of the user that is currently
logged in.

One or more network computers
Recommended for administrators only. Copies settings to the
LOCAL MACHINE registry of selected computers. Settings then
apply for new SolidWorks users on those machines.

One or more roaming user profiles
Recommended for administrators only. Copies settings to the
CURRENT USER registry of selected users. Select this option
only if your company uses roaming profiles.

In the **SolidWorks Copy Settings Wizard** dialog box, make sure that **Create backup of current settings for User** is not checked, and then click **Finish**. Since this **Restore** operation is just to get your SolidWorks back to when you started, you don't need to save the current settings because you don't plan to reuse these settings. If you wanted to save the current settings so that you can use them at a future time, either do a **Save Settings** first or check the **Create backup of current settings for User** check box.

Finish Operation

Click Finish to copy the settings to the current user, User.

☐ Create backup of current settings for User.
The backup will be stored as a .sldreg file in the folder chosen earlier.

A dialog box will appear confirming that the settings have been copied successfully. Click **OK**.

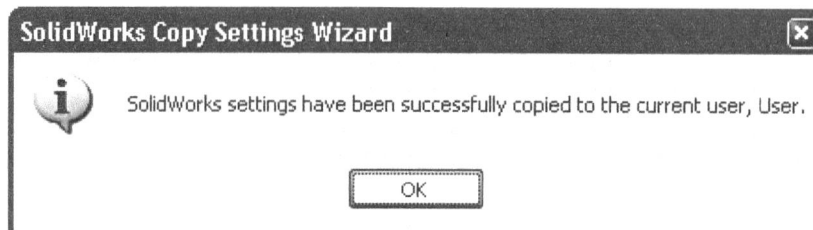

SolidWorks Copy Settings Wizard	☒
ⓘ SolidWorks settings have been successfully copied to the current user, User.	
OK	

Chapter 2

The user interface refers to the location of all the toolbars and windows. The actual geometry of the document is separate. SolidWorks allows you to customize the interface. The interface is simply the tools that you use and where and how you access them. You can make SolidWorks look any way that you like. Since the actual part data is not affected by the interface, multiple personal configurations may be used, which is ideal when more than one user shares a workstation. The interface background is fully customizable as well, including colors, skins, and background images.

The toolbar buttons are shortcuts for commands that you use frequently. You can customize the location and visibility of the toolbars based on the document type. The toolbar button size and the tooltip display can also be controlled. Each time a certain document type, part, assembly, or drawing, is opened, SolidWorks will display the toolbars where and how you placed them for that document type.

Reset All Preferences to Factory Defaults

Since all of your settings were saved in the previous chapter, you are now going to reset everything back to the default values and settings. This way, your SolidWorks will be set up similar to what you will see in this book, allowing you to instantly apply what you learn.

To begin, open the SolidWorks program.

Click the **Options** button in the "Standard" toolbar, or pull down the "Tools" menu and pick **Options**.

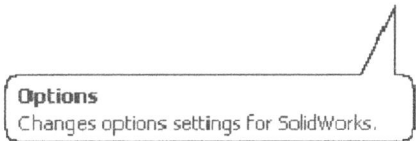

Options
Changes options settings for SolidWorks.

The **System Options** dialog box appears.

At the bottom of the **System Options** tab, click the **Reset All** button as shown to return all the system options to the system defaults.

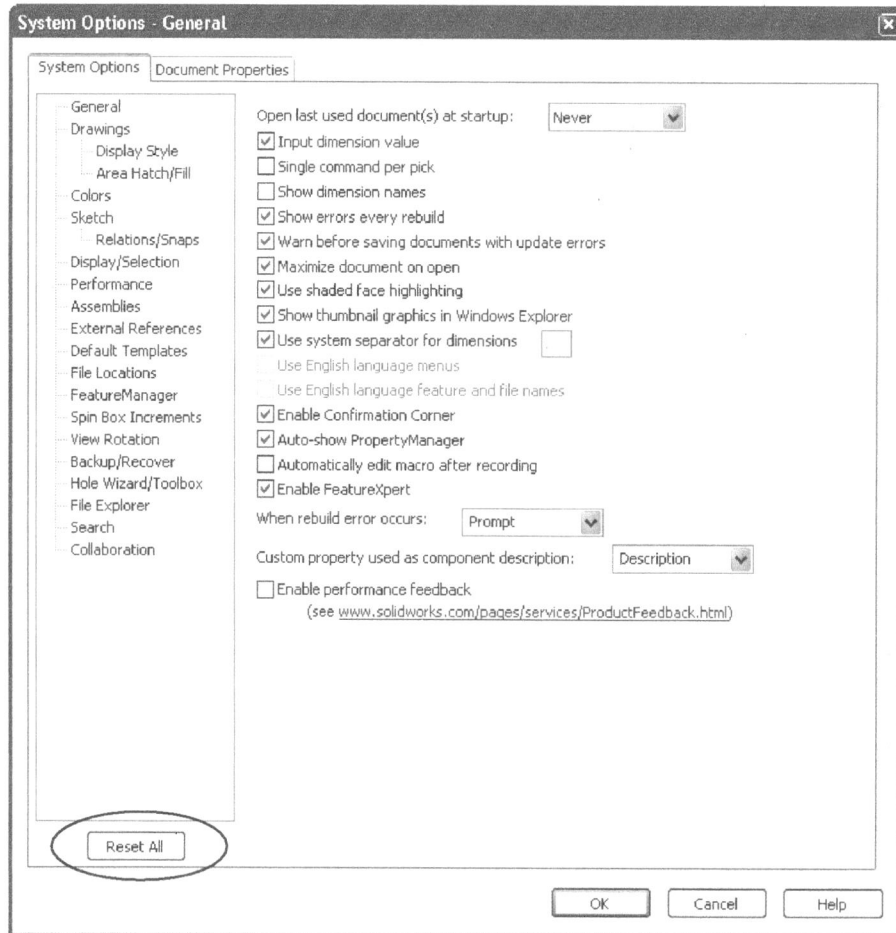

Click **Yes** in the **SolidWorks** dialog box to reset all preferences to the factory defaults, and then click **OK** in the **System Options** dialog box.

In the SolidWorks window, notice the toolbars at the top of the screen and on the right side of your screen.

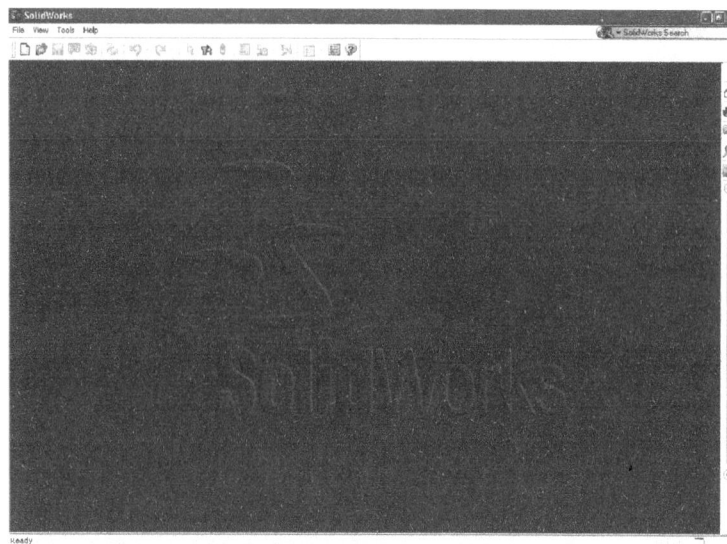

Customizing SolidWorks For Greater Productivity

You can customize the position and visibility of the toolbars based on the open document type or when no documents are open. If you pull down the "Tools" menu, you will notice that **Customize** is grayed out.

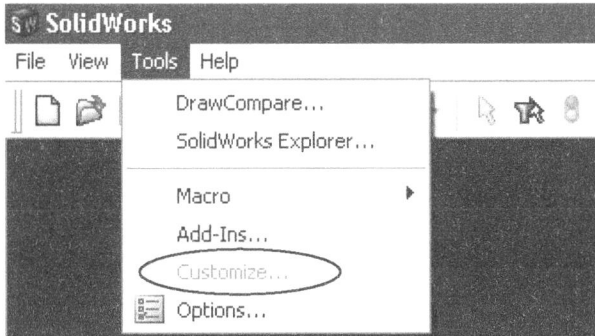

To indicate which toolbars are shown when no documents are open, move the cursor over the toolbars and press the right mouse button.

A menu appears showing all the available toolbars. The displayed toolbars are indicated by the selected button to the left of the toolbar name. ("Standard" and "Task Pane")

For now, do not make any changes. Simply click anywhere outside of the menu, or press the **Escape** key. This closes the menu.

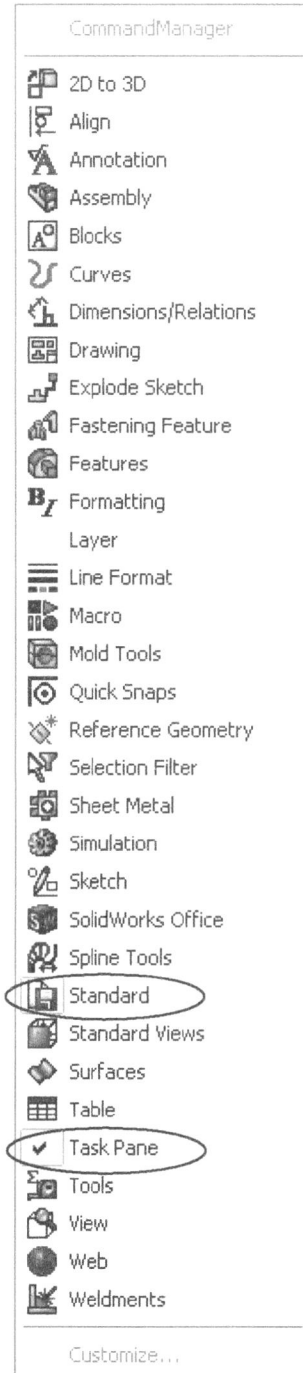

Create a New Part Document

Click the **New** button in the "Standard" toolbar, or pull down the "File" menu and pick **New**.

New
Creates a new document.

The **New SolidWorks Document** dialog box appears.

Click **Part** and then click **OK**. A new part window appears.

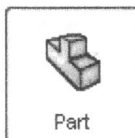

a 3D representation of a single design component

Set the Background Color

SolidWorks gives you the ability to change the look of the background of the graphics area and PropertyManager, allowing you to show a little bit of your personality in your working environment. SolidWorks uses different system colors throughout the program to provide visual feedback, like highlighting selected geometry or showing you that a sketch is over defined. The default colors are very valuable in being efficient at using the program. Even though you are able to change the color settings, it is recommended that you only change the background and PropertyManager appearance unless you have a specific reason for changing those other items.

To set the system colors, click the **Options** button in the "Standard" toolbar, or pull down the "Tools" menu and pick **Options**.

In the **System Options** dialog box, on the **System Options** tab, click **Colors** in the box on the left.

Options
Changes options settings for SolidWorks.

Under **Color scheme settings**, make sure that **Viewport Background** is selected. Then, click the **Edit** button to the right.

In the **Color** dialog box, under **Basic colors**, click on **Black**, and then click the **OK** button.

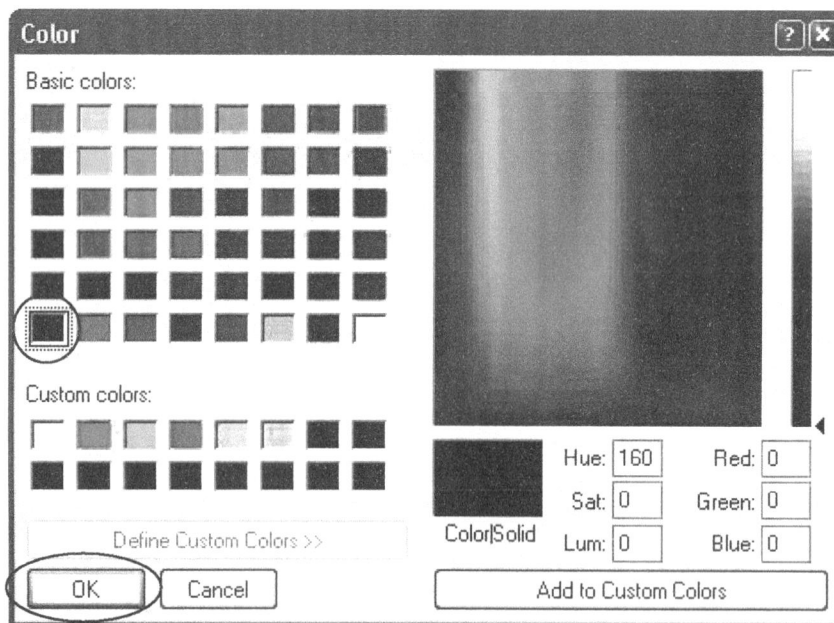

In the **System Options** dialog box, change the **PropertyManager color** by pulling down the menu and picking **Sand**.

Set the **Background appearance** by clicking the **Plain (Viewport Background color above)** radio button.

Uncheck **Match graphics area and FeatureManager backgrounds**, and click **OK**.

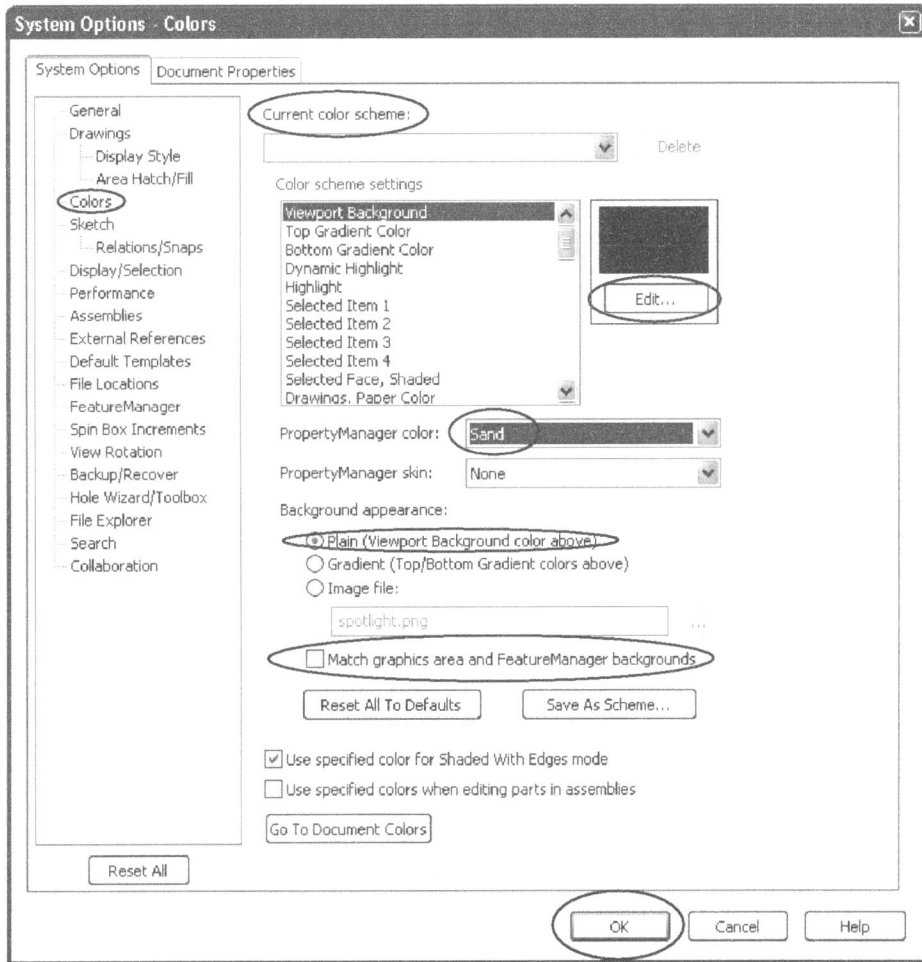

You can easily notice the black background and the sand colored PropertyManager.

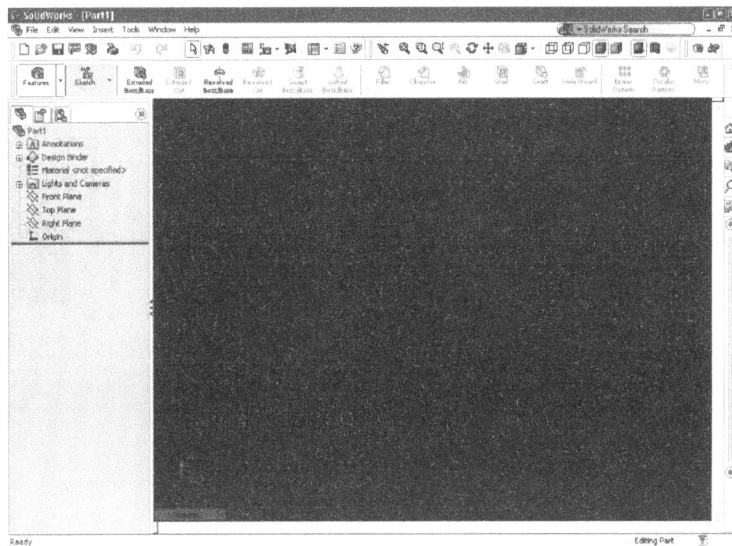

Chapter 2: User Interface Look and Feel   9

Create a Custom Color Scheme

A color scheme is an easy way to save numerous **Color scheme settings** in one place. This way, you can easily change from one set of colors to another.

To create your own color scheme, click the **Options** button in the "Standard" toolbar, or pull down the "Tools" menu and pick **Options**.

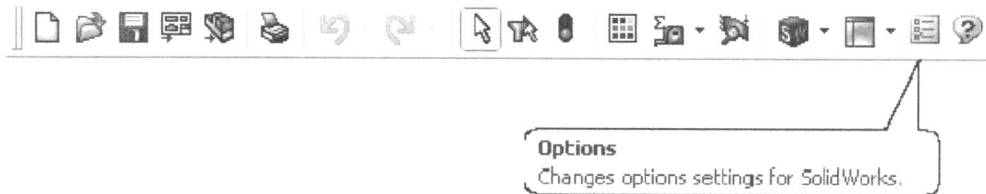

Options
Changes options settings for SolidWorks.

In the **System Options** dialog box, on the **System Options** tab, click **Colors**.

Any changes to the **Color scheme settings** can be saved. To save the current Sand / Black color scheme, under **Color scheme settings**, click the **Save As Scheme...** button.

Save As Scheme...

In the **Color Scheme Name** dialog box, type '**Sand / Black**', and click **OK**.

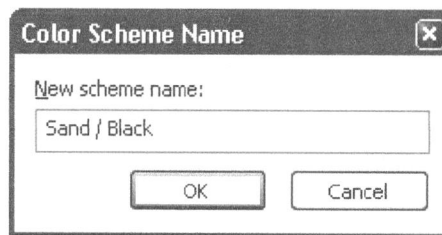

Color Scheme Name

New scheme name:

Sand / Black

OK   Cancel

The new color scheme appears in the **Current color scheme** pull down menu.

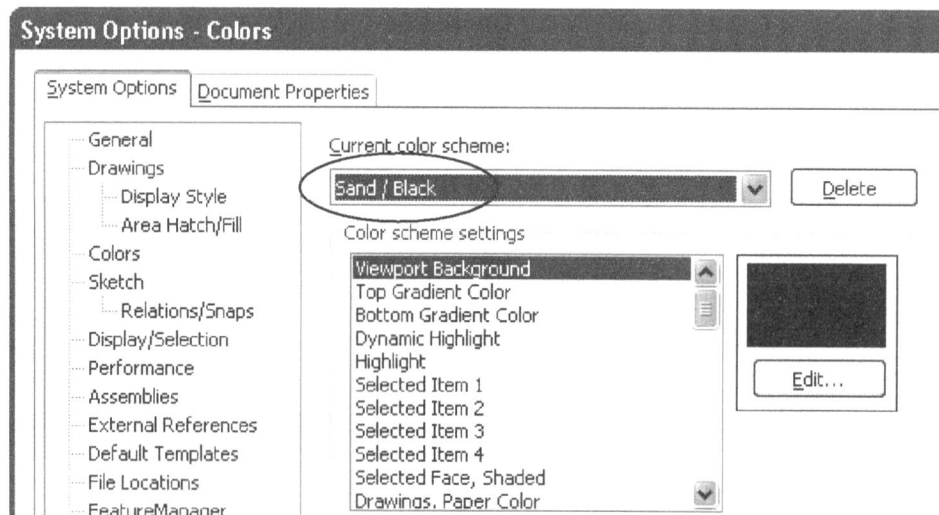

System Options - Colors

System Options | Document Properties

- General
- Drawings
  - Display Style
  - Area Hatch/Fill
- Colors
- Sketch
  - Relations/Snaps
- Display/Selection
- Performance
- Assemblies
- External References
- Default Templates
- File Locations
- FeatureManager

Current color scheme:

Sand / Black   Delete

Color scheme settings

Viewport Background
Top Gradient Color
Bottom Gradient Color
Dynamic Highlight
Highlight
Selected Item 1
Selected Item 2
Selected Item 3
Selected Item 4
Selected Face, Shaded
Drawings, Paper Color

Edit...

SolidWorks supplies some sample color schemes (shown below) to choose from. By default, the **Current color scheme** is set to **In the Spotlight**. The sample color schemes use different background colors, including a gradient background and even an image file for the background.

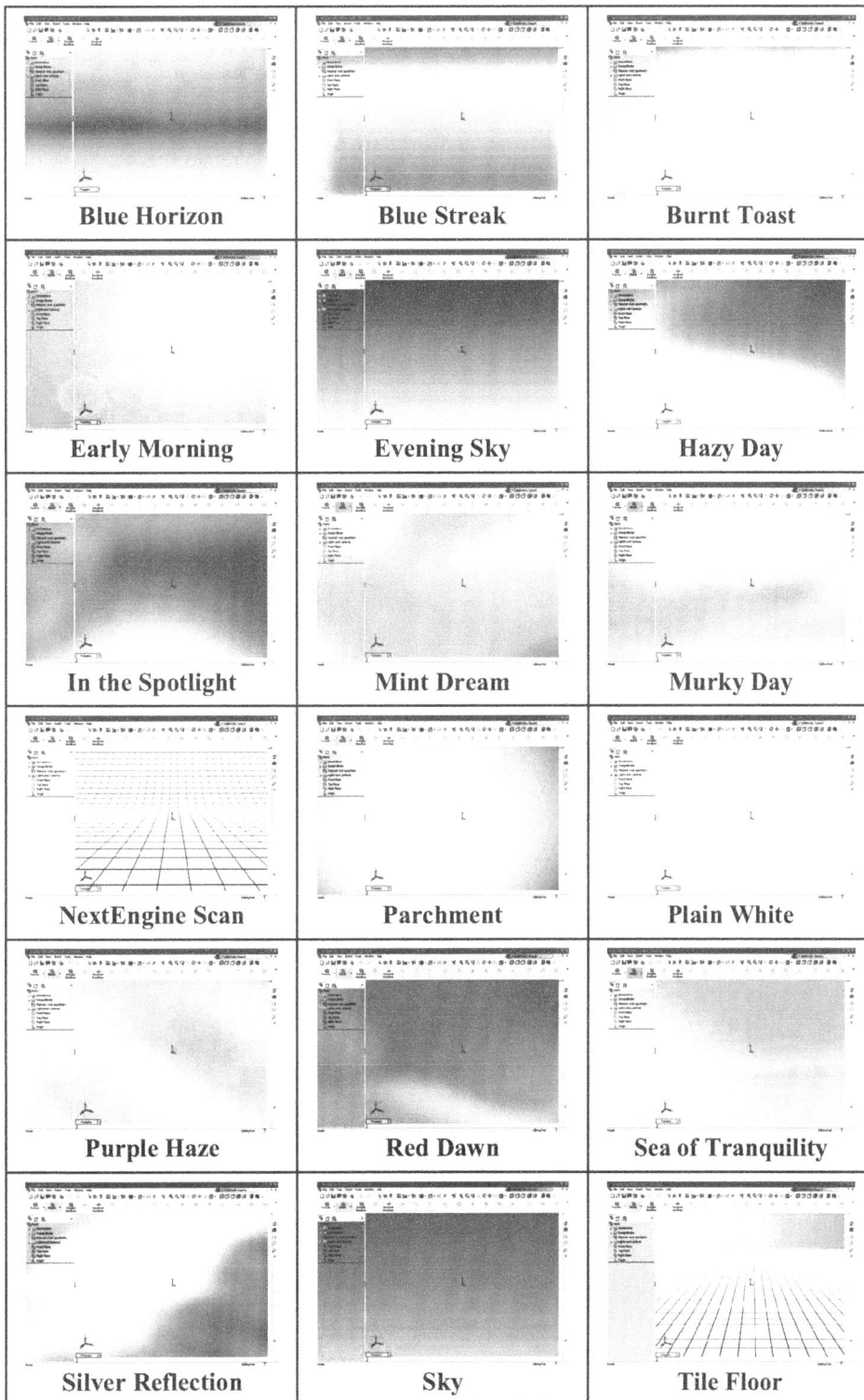

Blue Horizon	**Blue Streak**	**Burnt Toast**
Early Morning	**Evening Sky**	**Hazy Day**
In the Spotlight	**Mint Dream**	**Murky Day**
NextEngine Scan	**Parchment**	**Plain White**
Purple Haze	**Red Dawn**	**Sea of Tranquility**
Silver Reflection	**Sky**	**Tile Floor**

Gradient Background Color

Besides a plain, one color background, you can customize the background to be a gradient vertical fade between two colors, from the top of the graphics area to the bottom.

Click the **Options** button in the "Standard" toolbar, or pull down the "Tools" menu and pick **Options**.

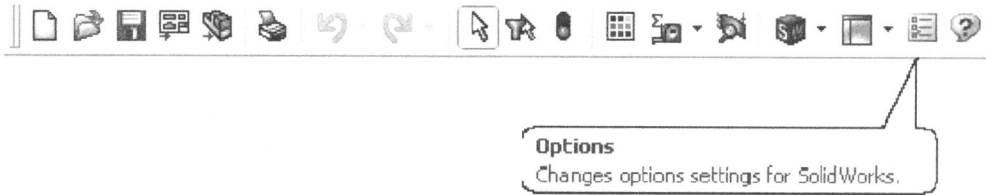

Options
Changes options settings for SolidWorks.

In the **System Options** dialog box, on the **System Options** tab, click **Colors**.

To change the background to a gradient, under **Color scheme settings**, edit **Top Gradient Color** and **Bottom Gradient Color**.

Then, under **Background appearance**, pick the **Gradient (Top/Bottom Gradient colors above)** radio button and click **OK**.

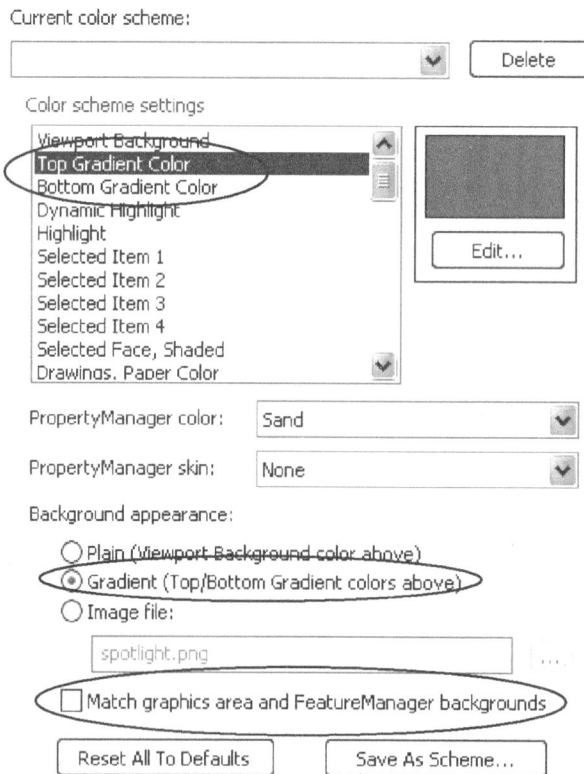

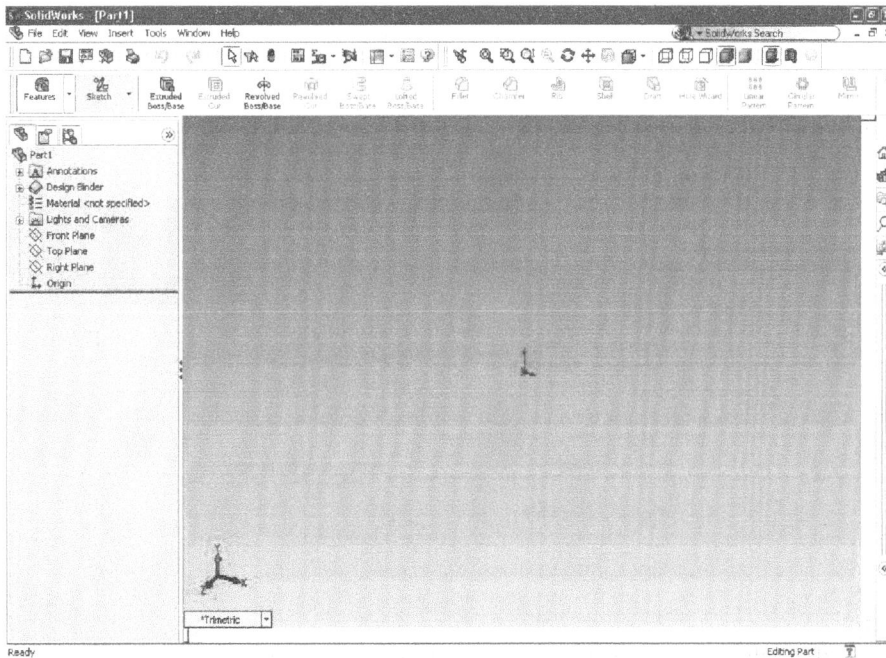

Note that since **Match graphics area and FeatureManager backgrounds** was not checked, the left panel is still Sand colored. Try some different color combinations to see what you like. You may want to save it as a color scheme.

Background Image File

Another way to add your personal touch to SolidWorks is to use an image file for your background. You may want to use a picture of your family, favorite car, vacation place, or company logo. Although this is not required to run SolidWorks, it does make customizing SolidWorks a lot of fun.

To do this, click the **Options** button in the "Standard" toolbar, or pull down the "Tools" menu and pick **Options**.

Options
Changes options settings for SolidWorks.

In the **System Options** dialog box, on the **System Options** tab, click **Colors**.

Under **Background appearance**, pick the **Image file** radio button.

Then, click ⌷ next to the image file name to browse for an image file. Just as you would change the Windows desktop background, SolidWorks lets you do so using any image file formats that you have.

Chapter 2: User Interface Look and Feel   13

Pick **studio2.jpg** from the background folder, <SolidWorks Install Directory>/data/images/textures/background, and click **Open**. You may also just double click on the file to open it.

Check **Match graphics area and FeatureManager backgrounds**, and click **OK**.

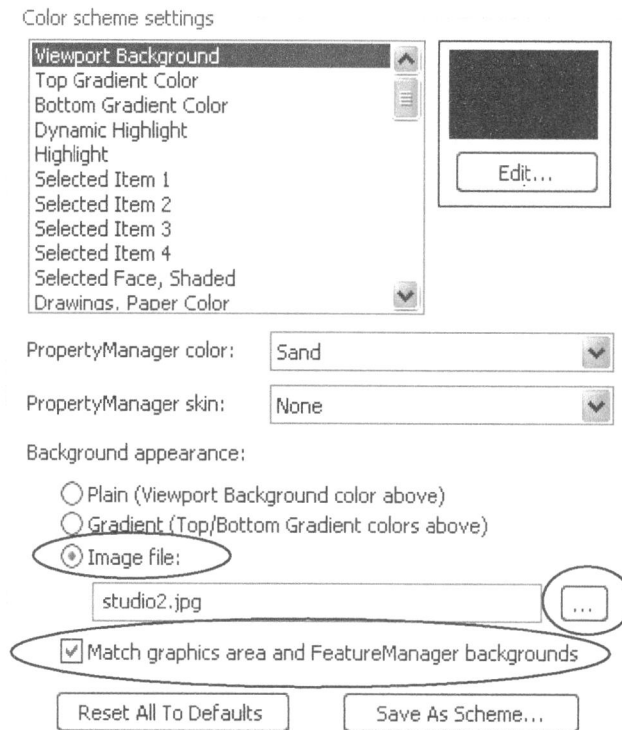

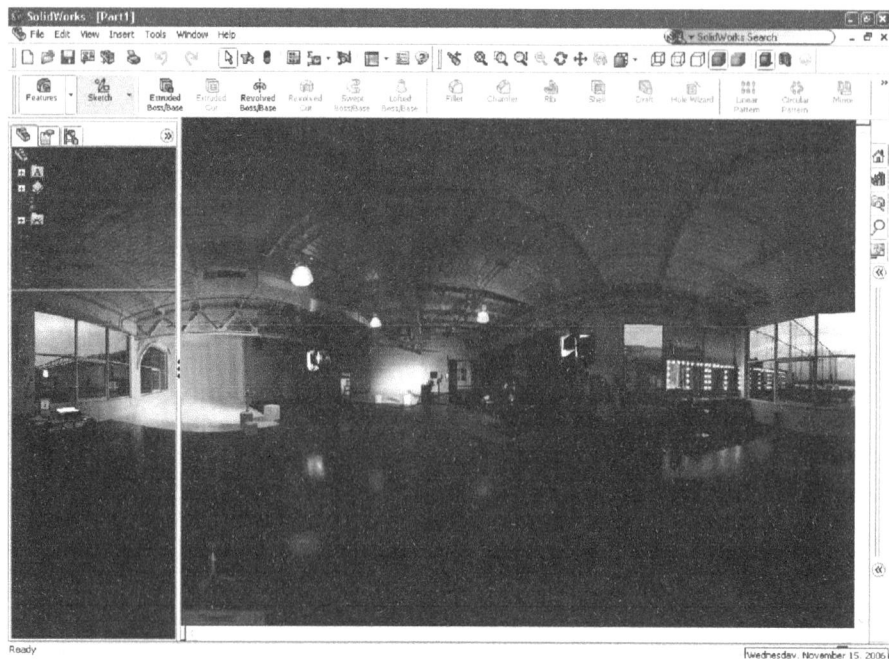

Note that if an image file is specified, the image file takes precedence over the other background color options.

Try out some other image files. These are three images I found on my computer. Have some fun!

PropertyManager Look and Feel

Another customizable feature that SolidWorks offers is PropertyManager skins. A skin is just a tiled bitmap image used as a background in the PropertyManager for more of a personal touch. Perhaps the image file behind the graphics area is just too much, but you still want a little bit of fun. Since skins are only seen in the PropertyManager, you will only see them while doing things such as editing the definition of a feature, or displaying the properties of an entity.

By default, the **PropertyManager skin** is set to **None**. SolidWorks supplies some sample skins to choose from (shown below).

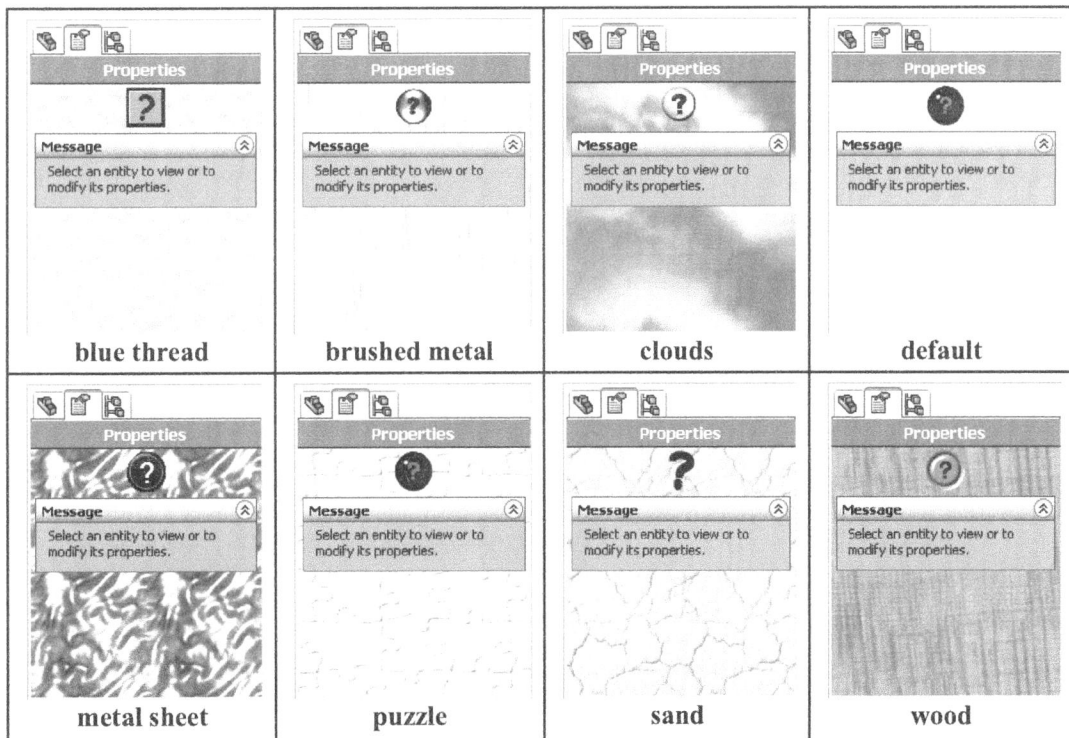

blue thread	**brushed metal**	**clouds**	**default**
metal sheet	**puzzle**	**sand**	**wood**

To apply a skin, click the **Options** button in the "Standard" toolbar, or pull down the "Tools" menu and pick **Options**.

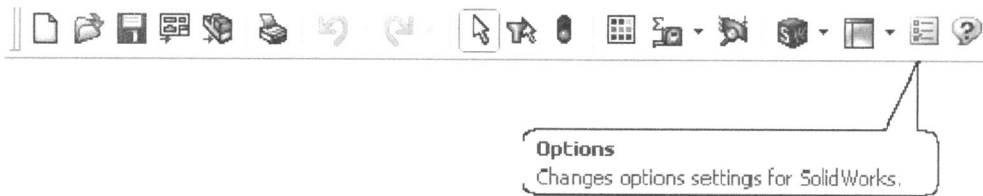

> **Options**
> Changes options settings for SolidWorks.

In the **System Options** dialog box, on the **System Options** tab, click **Colors**.

Pull down the **Current color scheme** menu and pick **Plain White**.

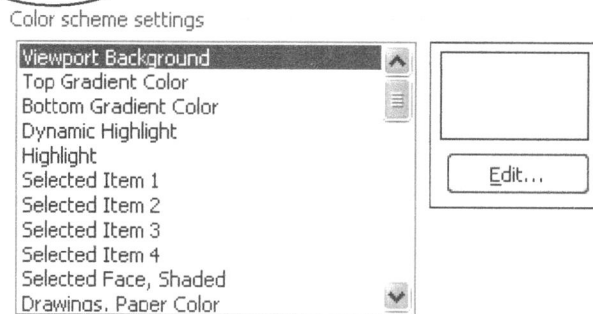

Notice in the dialog box that the **Viewport Background** color changed. The **PropertyManager color** changed to **Windows**, and the **Background appearance** changed to **Plain**.

Under **Color scheme settings**, pull down the **PropertyManager skin** menu and pick **clouds**.

In the **System Options** dialog box, click the **OK** button.

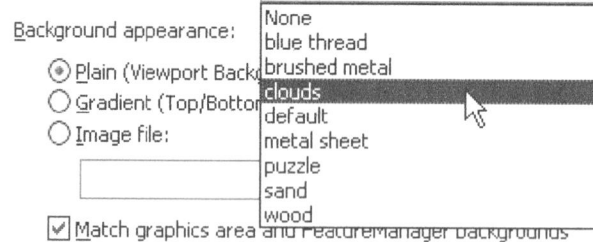

In the FeatureManager tree area, click on the **PropertyManager** tab.

Only the background of the PropertyManager is now clouds.

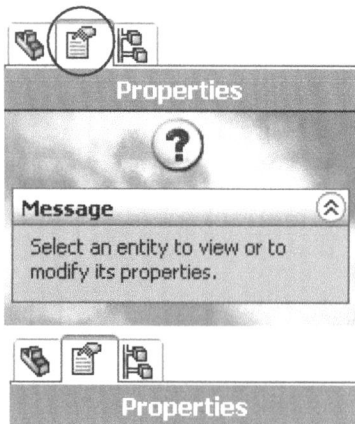

Current color scheme:

Plain White   Delete

Color scheme settings

Viewport Background
Top Gradient Color
Bottom Gradient Color
Dynamic Highlight
Highlight
Selected Item 1
Selected Item 2
Selected Item 3
Selected Item 4
Selected Face, Shaded
Drawings, Paper Color

Edit...

PropertyManager color: Windows

PropertyManager skin: None

Background appearance:

None
blue thread
brushed metal
clouds
default
metal sheet
puzzle
sand
wood

○ Plain (Viewport Back...
○ Gradient (Top/Botto...
○ Image file:

☑ Match graphics area and FeatureManager backgrounds

Reset All To Defaults   Save As Scheme...

Properties

(?)

Message   (⊗)

Select an entity to view or to modify its properties.

Properties

(?)

Click the **Options** button in the "Standard" toolbar, or pull down the "Tools" menu and pick **Options**.

Under **Color scheme settings**, pull down the **PropertyManager skin** menu and pick **None**.

In the **System Options** dialog box, click the **OK** button.

The background of the PropertyManager is now white again.

You can create your own custom skins. SolidWorks requires the skin image to be a bitmap file. The bitmap (.bmp) files are saved in the skins folder, <SolidWorks Install Directory>/data/skins. When editing or creating a bitmap image, consider its size. The width of the FeatureManager tree area snaps to about 190 pixels, so that is a good size to make the skin image. Ideally, the image file selected will tile well so that it makes a seamless background. Use your favorite image program, like Microsoft Paint or Adobe Photoshop, to create the bitmap. Once a bitmap is ready for use, save it in the SolidWorks Install Directory/data/skins folder.

Besides the background image of a skin, you can also create your own PropertyManager customized buttons. The easiest way to create your own buttons is just to copy an existing skin and modify the bitmap image copies. To do this, copy similar buttons and rename the files in Windows Explorer with a new skin name. It is required that the file name have the exact skin name, followed by two underscores, then the button function (cancel, help, ok, pin), and then the file extension. Otherwise, the default buttons will be used. For example, if the skin is named **skin.bmp,** the cancel button for that skin would be named **skin__cancel.bmp.** Customize each button to your liking and save the bitmaps using the new skin name. Make sure that each button is renamed properly.

Once all the files are saved in the skins folder, the skin will appear in the PropertyManager skin pull down menu for your use. More skins are available at www.SheetMetalGuy.com/solidworks.

Notice in the "Standard" toolbar, there is a button, **Select Color Scheme,** which rotates through the list. Or you can use the arrow to pull down the list and select the one you want.

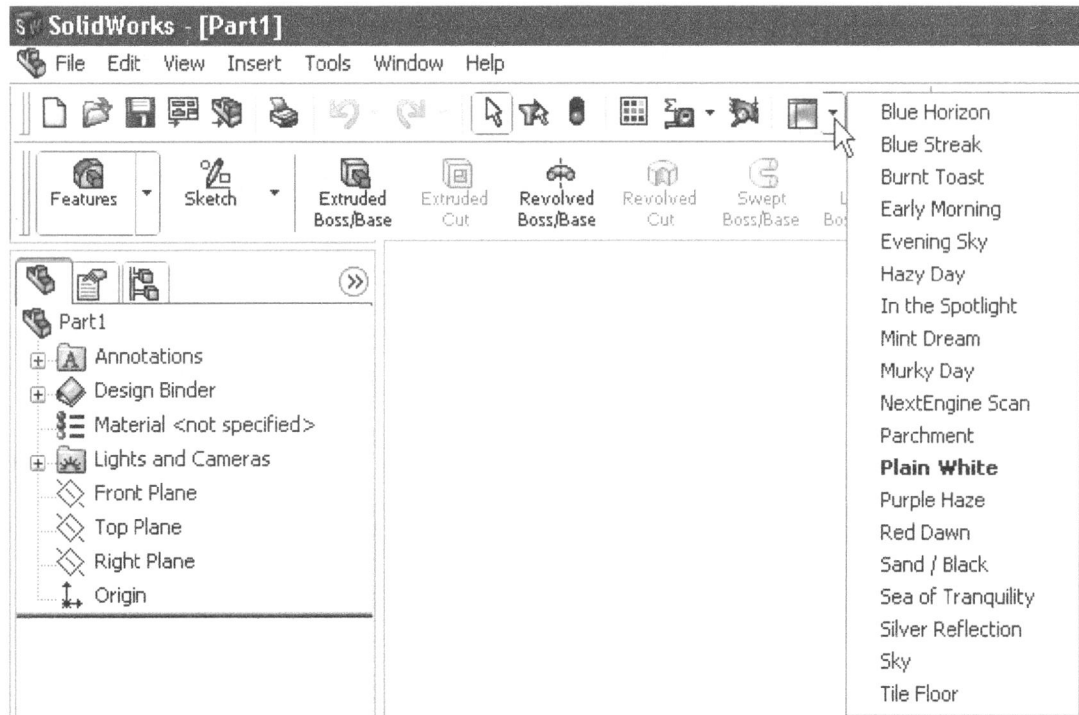

Toolbar Location and Display

Since a **Part** document is open, any changes to the position and visibility of the toolbars will only affect part files. Your selections apply only to the type of SolidWorks document that is currently active. To customize the toolbars of your drawing or assembly documents, open a new drawing or assembly document and customize the position and visibility of the toolbars for each file type.

To begin, move the cursor over a toolbar and press the right mouse button. Make sure that you don't right click over the CommandManager.

Pick **Standard Views** from the menu. The "Standard Views" toolbar appears docked at the top of the upper right hand corner of the SolidWorks window.

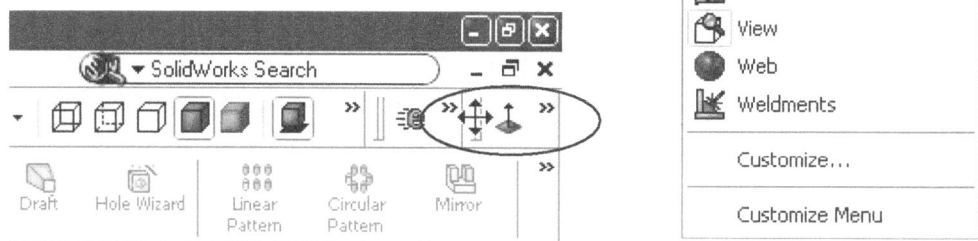

To undock the "Standard Views" toolbar and move it to a new location, place the cursor at the start of the toolbar. Each docked toolbar will have two lines either at the start or the top of the toolbar, depending on the toolbar orientation. When the cursor is over the two lines, it will change from a single arrow cursor to a four-way arrow cursor, as shown above.

Do it with me! Hold down the left mouse button and drag the "Standard Views" toolbar by moving the mouse. Drag the toolbar into the middle of the graphics area and release the mouse button. The toolbar is left in place floating (undocked). If you drag the toolbar to any side of the screen, the outline becomes narrower. When you release the left mouse button, the toolbar becomes docked. You can place the toolbars wherever you choose on the screen by docking them where you like. Toolbars can be docked horizontally or vertically, on the top, bottom, left, or right side of the screen.

Now try this. Double click on the header of the "Standard Views" toolbar. Double clicking the start or title bar of a toolbar moves the toolbar back to its previous position. Pretty Cool!

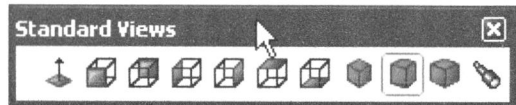

Place the cursor at the start of the toolbar again. This time, hold down the **Ctrl** key and drag the "Standard Views" toolbar on top of the right side of the CommandManager. Release the mouse button and then the **Ctrl** key. Holding down the **Ctrl** key prevents the toolbar from docking.

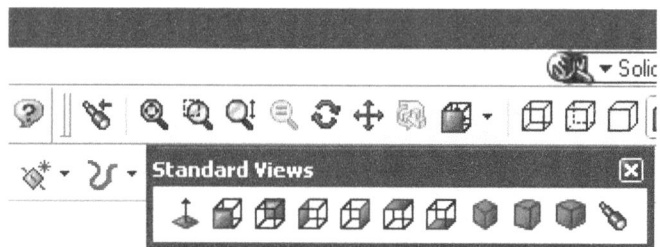

Display of All Toolbars

Rather than changing the visibility of the toolbars one by one, you can select the toolbars you want to display and clear the toolbars you want to hide all at the same time.

To do this, right click over one of the toolbars, and pick **Customize**, or pull down the "Tools" menu and pick **Customize**. You may need to scroll the menu down to see it.

In the **Customize** dialog box, on the **Toolbars** tab, scroll down until you see **Standard Views**.

Uncheck **Standard Views** and click **OK**. You could check or uncheck others while you are here.

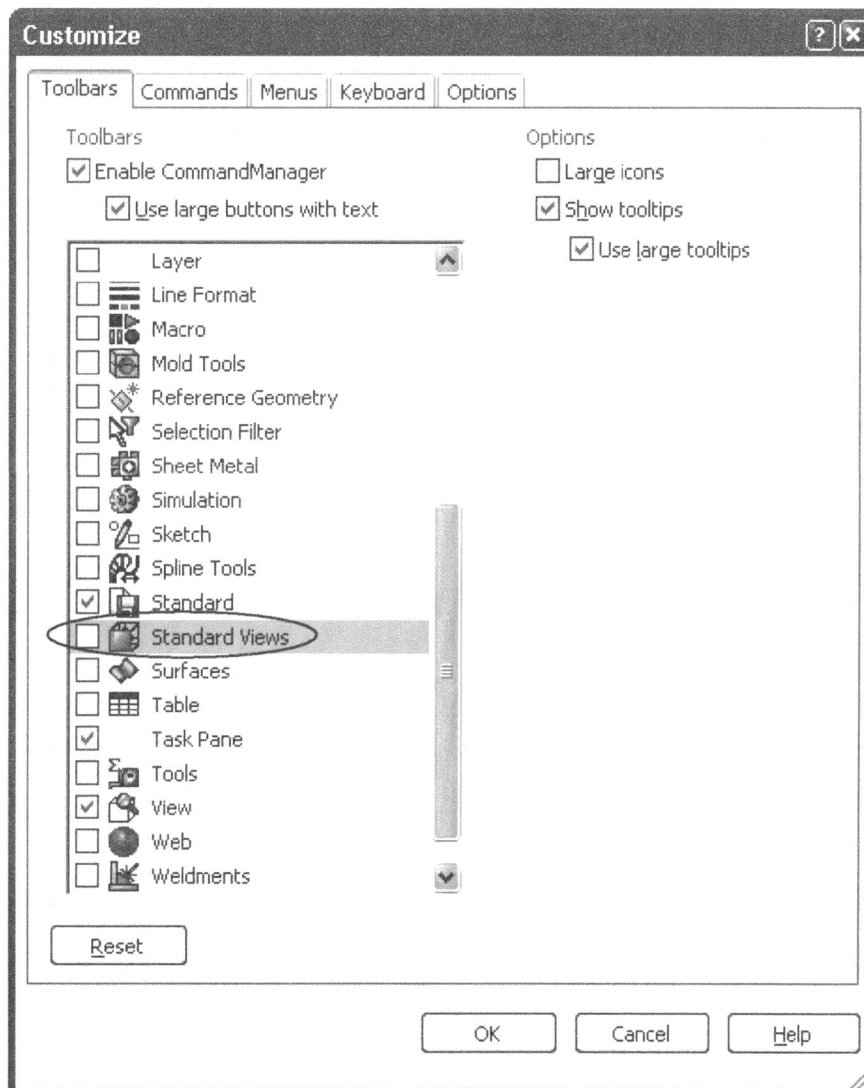

Another way to change the display of a toolbar is to pull down the "View" menu and pick **Toolbars**. Click on the desired toolbar to toggle its display. The method shown above is best when you need to change the display of several toolbars at the same time.

All the standard toolbars are listed in the **Customize** dialog box. To change the display of a SolidWorks Add-In toolbar, you can access the toolbar from the right click menu. Add-Ins are not included in the **Customize** dialog box.

To change the display of the "eDrawings 2007" toolbar in the upper right-hand corner of the SolidWorks window, right click over the toolbars and uncheck **eDrawings 2007** to turn the display of the toolbar off. Check **eDrawings 2007** to turn the display of the toolbar on. When a toolbar is turned off, then turned back on, it reappears in its most recent position.

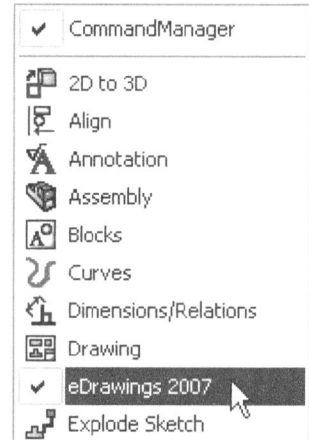

To temporarily turn off all of the toolbars, press the **F10** key, or pull down the "View" menu and pick **Toolbars** (not the pull out menu, look further down in the menu).

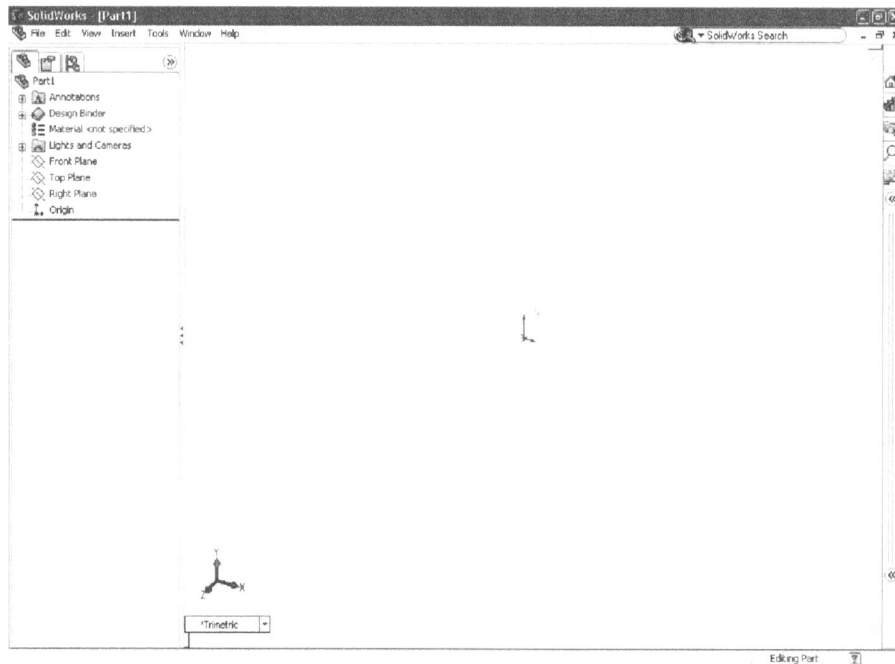

Press the **F10** key again, or pull down the "View" menu and pick **Toolbars** to toggle back on the display of all the active toolbars.

Icon Size and Tooltip Display Options

To change the settings for icon size and tooltips, pull down the "Tools" menu and pick **Customize**, or right click on a toolbar and pick **Customize** from the bottom of the menu.

In the **Customize** dialog box, on the **Toolbars** tab, there are some check boxes in the upper right corner. Check the **Large icons** check box.

Make sure that **Show tooltips** and **Use large tooltips** are checked, and then click **OK**. Note that **Use large tooltips** is only available when **Show tooltips** is checked.

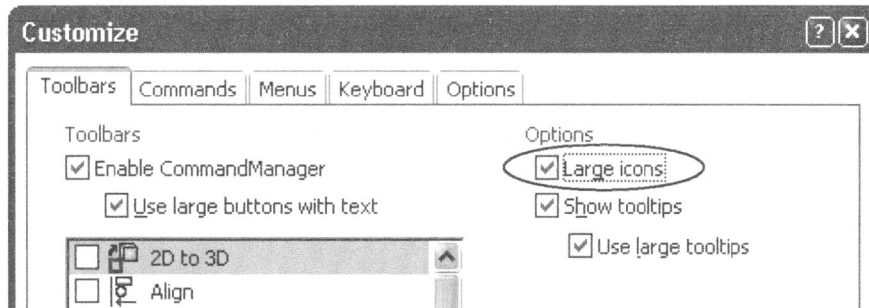

The toolbars and buttons are now much larger in the SolidWorks window. When you place the cursor over the toolbar buttons, tooltips appear that display the name of the tool button and a brief description of its functionality. The large buttons are great when you have a very high resolution monitor.

Pull down the "Tools" menu and pick **Customize**, or right click on a toolbar and pick **Customize**.

Uncheck **Large icons** and **Use large tooltips**, and then click **OK**.

Now, when you place the cursor over a toolbar, a small tooltip appears that describes the command that the cursor is over. Note that the large tooltips usually include more information.

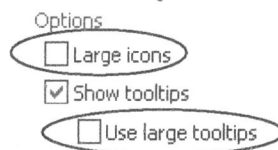

Pull down the "Tools" menu and pick **Customize**, or right click on a toolbar and pick **Customize**.

Check **Use large tooltips**, and then click **OK**.

Left Panel Display

The left panel, by default, includes the FeatureManager design tree, PropertyManager, and ConfigurationManager. You can hide or show this left pane.

Press the **F9** key to hide the left panel, or click the arrow button in the center of the panel border, or pull down the "View" menu and pick **FeatureManager Tree Area**.

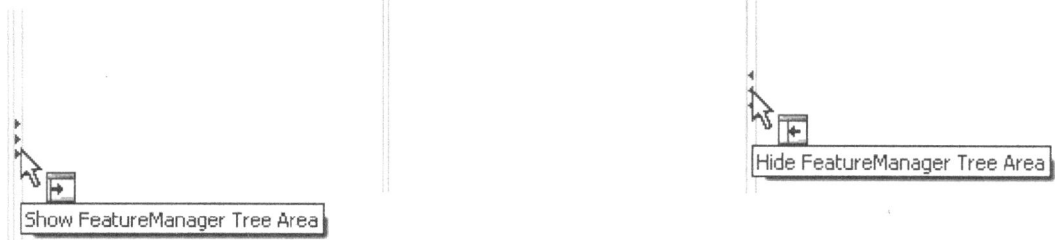

Press the **F9** key again to show the left panel, or click the arrow button in the center of the panel border, or pull down the "View" menu and pick **FeatureManager Tree Area**.

You can also change the size of the left panel at any time. To do this, move the cursor over the panel border. When the cursor changes as shown, press the left mouse button and drag the border to the desired size.

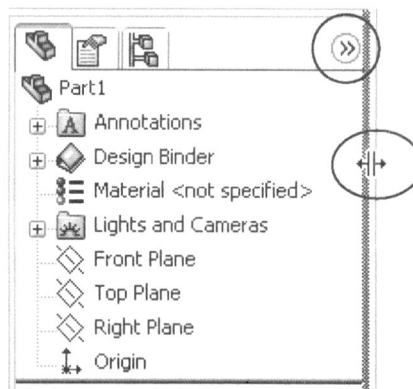

Click the arrow in the upper right corner of the left panel. This shows the Display Pane. In the Display Pane, you can apply the following display settings for each component in an assembly: Hide/Show state, Display mode, Component color, Component texture, and Transparency. In part and drawing documents, you can view this information, but not modify it.

⊗ Click the arrow to collapse the Display
Pane.

Task Pane Location and Display

The Task Pane allows you to access valuable resources, design libraries, and other files and information. By default, the Task Pane appears on the right hand side of the SolidWorks window.

⊗ Expand the Task Pane by clicking an arrow or anywhere along the double vertical bar between the arrows.

⊗ Collapse the Task Pane by clicking anywhere in the graphics area or by clicking an arrow or anywhere along the double vertical bar between the arrows.

⊟ The task Pane closes whenever you click the mouse or perform a command outside of the Task Pane. To keep the Task Pane expanded, expand the Task Pane and click the **Pin** button in the title bar.

Now, when you click anywhere in the graphics area, the Task Pane remains expanded. It does not collapse since it is pinned.

⊡ Click the **Pin** button again to unpin the Task Pane, allowing it to collapse.

To float the Task Pane, drag it by the bar between the arrows into the graphics area.

You can resize the floating Task Pane by dragging any edge that is not docked.

Collapse the Task Pane vertically by clicking anywhere in the graphics area or by clicking the up arrow in the upper right corner of the Task Pane.

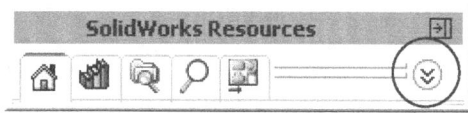

Expand the Task Pane vertically by clicking the down arrow or anywhere along the bar.

Move the Task Pane around by clicking in the title bar and dragging it around. Just like toolbars, holding down the **Ctrl** key when dragging prevents the Task Pane from docking.

Click the **Pin** button in the title bar to keep the Task Pane expanded. Again, the Task Pane will not collapse if it is pinned.

Click the **Pin** button to unpin the Task Pane, allowing it to collapse.

Dock the Task Pane by clicking the **Dock Task Pane** button in the upper right corner of the Task Pane.

Now try this. Double click on the title bar of the Task Pane. The Task Pane will move back to its floating position. Double click the title bar of the Task Pane again to re-dock the Task Pane.

Drag the Task Pane by the bar between the arrows to the left side of the screen. The Task Pane will dock to the left side.

When expanded, the Task Pane will open over the FeatureManager tree area.

When collapsed, the Task Pane will be seen to the left of the FeatureManager tree area.

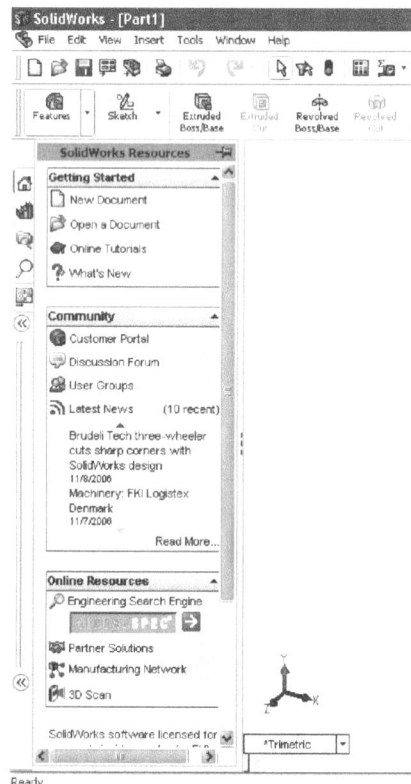

Drag the Task Pane back into the graphics area so that it is floating. Note that the **Dock Task Pane** button has now switched to dock the Task Pane to the left side.

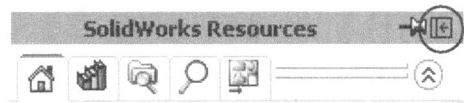

Drag the Task Pane by the title bar back to the right side to re-dock it to the right side.

Just like customizing the display of the toolbars, right click on the Task Pane bar and pick **Task Pane** from the menu to hide it. Click it again to show it. You can also pull down the "View" menu and pick **Task Pane** to do the same thing.

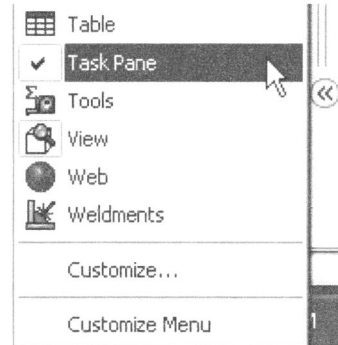

Full Screen Mode

You can maximize your drawing area by working in SolidWorks with the menus, status bar, and the FeatureManager tree area hidden.

To do this, press the **F11** key, or pull down the "View" menu and pick **Full Screen**. This hides the Windows Start bar, the Task Pane, and the toolbars.

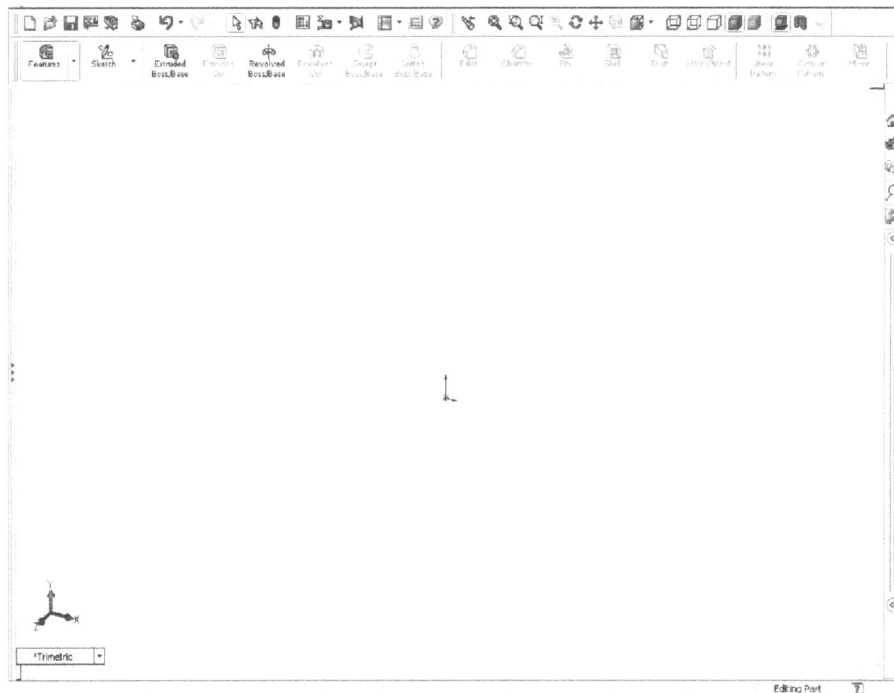

To hide or show the FeatureManager tree area, press the **F9** key.

To hide or show the toolbars, press the **F10** key.

To hide or show the Task Pane, pull down the "View" menu and pick **Task Pane**. Yes, you can still access the menus while in full screen mode. Just move the cursor to the top of the screen to pick the desired menu. Even though you can't see them, they are there at the very top edge of the screen. Look closely and click to open a menu then moving the mouse left or right displays the menu the cursor is under.

Note that the hide/show status for the Full Screen Mode is remembered separately from the Normal mode. In other words, if you hide the toolbars in the Full Screen Mode, when you return to Normal mode the toolbars will not be hidden.

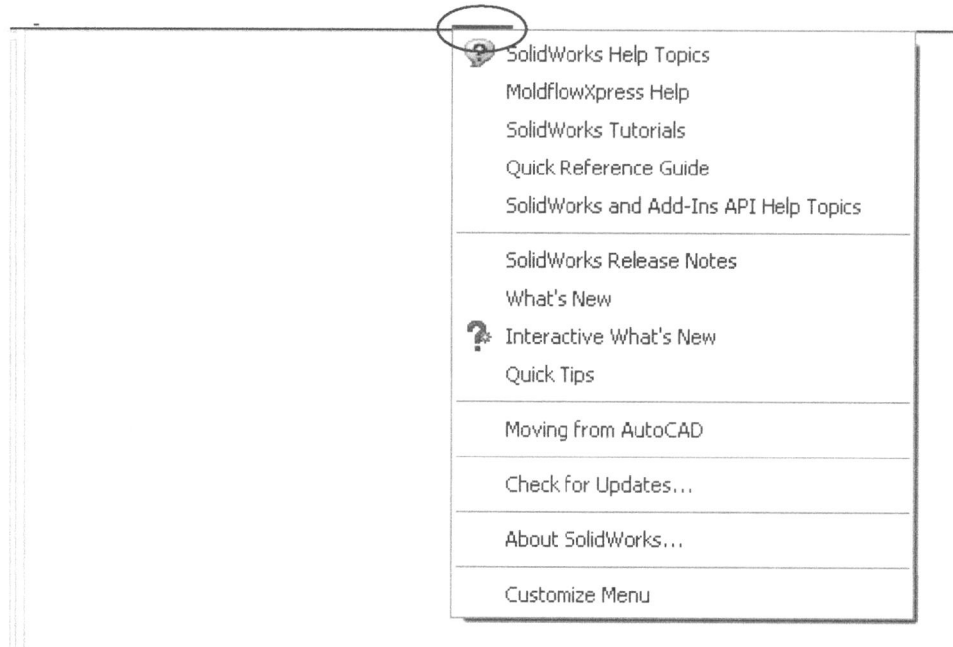

Press the **F11** key again, or pull down the "View" menu and pick **Full Screen** to return to the Normal mode.

Chapter 3

Customize Your Menus

There are many ways to increase your efficiency and effectiveness using SolidWorks. One of those ways is to customize your menus. There are many different ways to access the same commands in SolidWorks, allowing you to choose which way is best for you. The biggest advantage to customizing your menus is that the tools you use most can be easily accessed, and the tools that you never use can be hidden away so they are not in your way.

With many programs, you must use the menus the developers provide, regardless of whether you think it would be better for a specific command to be in different menu. SolidWorks allows you to modify the program, creating your own ideal set of menus by customizing which commands appear in which menus. One disadvantage is that functions may be buried in submenus, requiring a few clicks to find the command. SolidWorks allows you to choose what you would like to see in the Menu Bar. The only thing that you can't change is the name of the pull down menu and the order of the pull down menus.

Create a New Part Document

Pull down the "File" menu and pick **New**.

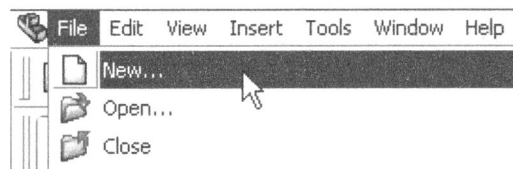

The **New SolidWorks Document** dialog box appears.

Click **Part** and then click **OK**. You may also just double click on **Part**.

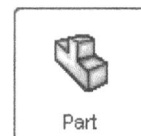

Part

Hide or Show Items in a Menu

Customizing the Menu Bar is pretty self-explanatory. At the bottom of each pull down and pull out menu is the **Customize Menu** command. Go ahead and pull down some menus to see for yourself. When this command is selected, you can customize what menu choices are shown in that pull down menu. This allows you to show only the items that you use most often without having to dig through a long list. Note that each document type (Part, Assembly, and Drawing) has its own menus. So, you must customize the menus for each document type.

To try this, pull down the "Help" menu and pick **Customize Menu**.

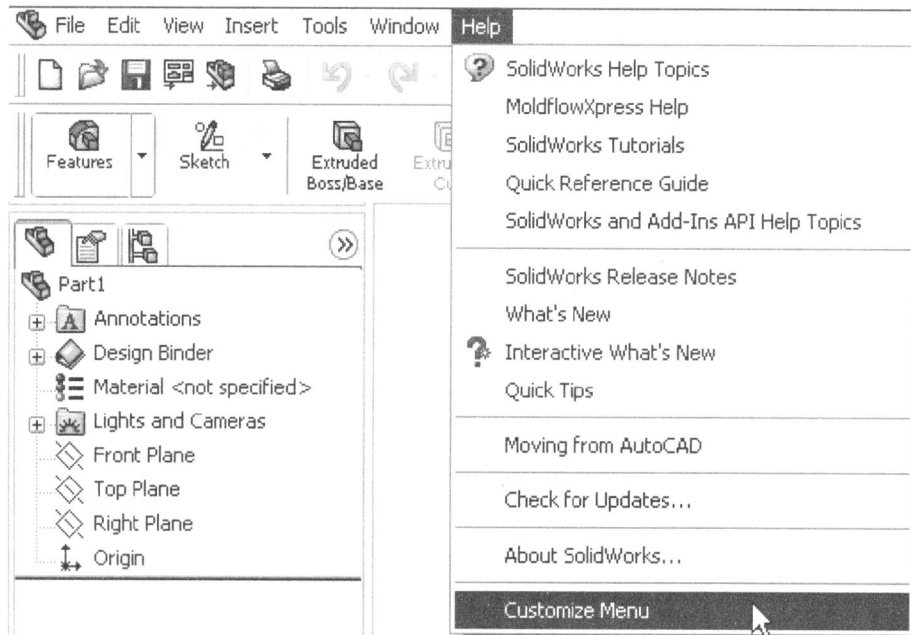

Uncheck **MoldflowXpress Help**, **Quick Reference Guide**, **SolidWorks and Add-Ins API Help Topics**, **SolidWorks Release Notes**, **What's New**, **Interactive What's New**, **Quick Tips**, and **Moving from AutoCAD** as shown. Check the check boxes to show a menu item or clear the check boxes to hide a menu item.

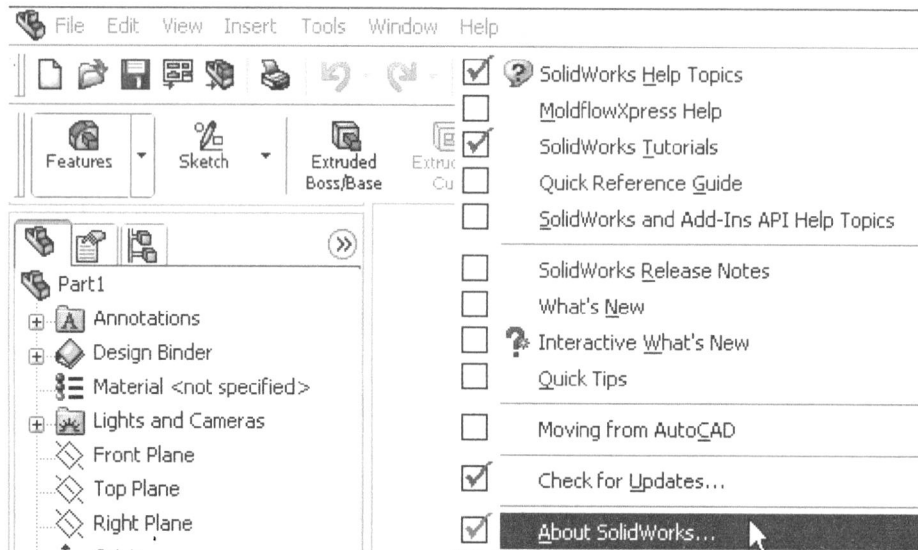

Click outside the menu or press **Enter** to accept the changes. Note that pressing the **Escape** key cancels your changes.

Now, when you pull down the "Help" menu, the list of choices is much shorter.

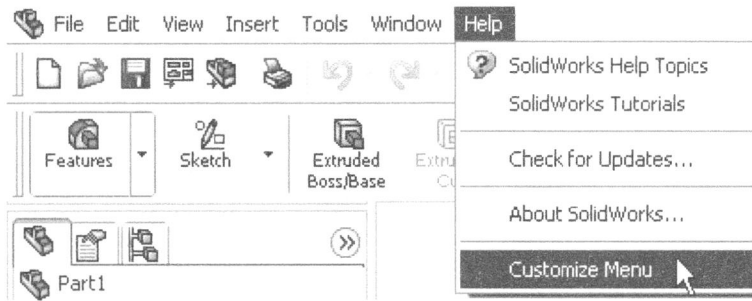

Another area like this is the "Insert" menu. If you don't work with **Sheet Metal**, **Weldments**, or **Molds**, you can customize this menu to turn off these items.

You can show or hide menu choices at any time by picking **Customize Menu**. Just remember that it is very easy to forget that a command is turned off.

Customize the Shortcut Menus

SolidWorks uses shortcut menus and right mouse button menus to help you quickly access common commands for what you are doing, or where your cursor is. Note that different shortcut menus appear depending on what command is active, where your cursor is located, or what is selected. Shortcut menus and right mouse button menus can be customized the same way the Menu Bar pull down menus are customized. To access the desired shortcut menu, you will need to start a command which offers that menu.

With no command active, right click in the graphics area to display the shortcut menu.

Click the down arrow to show the full menu, and then pick **Customize Menu**.

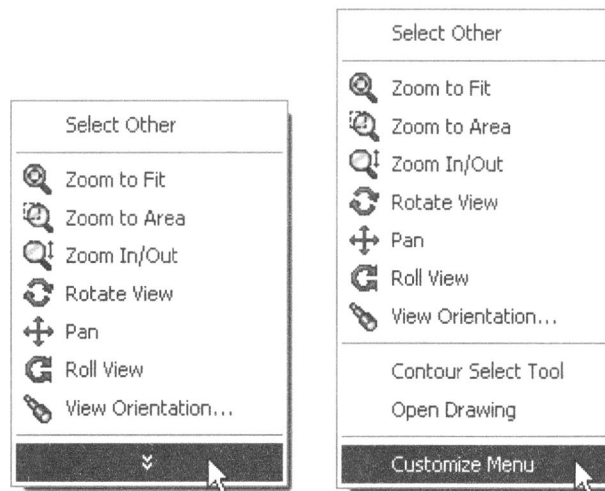

Note that only the selected items appear on the short version of the shortcut menu. When you click on the down arrow, the long version of the shortcut menu is shown, which shows all selected and cleared items.

Uncheck **View Orientation** as shown. Check the check box to show a menu item or clear the check box to hide a menu item.

Click outside the menu or press **Enter**. Note that pressing the **Escape** key cancels your changes.

Now, when you right click in the graphics area to display the shortcut menu, the list of choices is shorter. **View Orientation** is not shown.

This is nice, because you can reduce these shortcut menus down to the few items you use all the time. Yet the full menu is available simply by clicking the down arrow at the bottom of the menu.

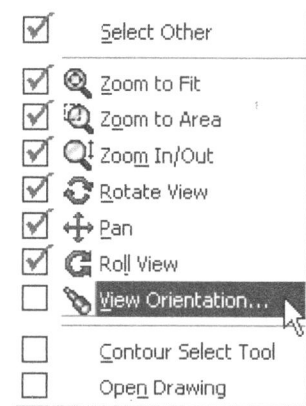

Show All and Reset to Defaults

By default, SolidWorks does not show all of the items available in each menu. You do have the ability to have SolidWorks show all of these items. You also have the ability to restore the items in the menus back to the default display.

Pull down the "Tools" menu and pick **Customize** or right-click over the toolbars and pick **Customize**.

In the **Customize** dialog box, click on the **Options** tab.

Under **Menu customization**, click the **Show All** button and then click **OK**.

For the shortcut menus, just use the **Shortcut customization** instead of **Menu customization**.

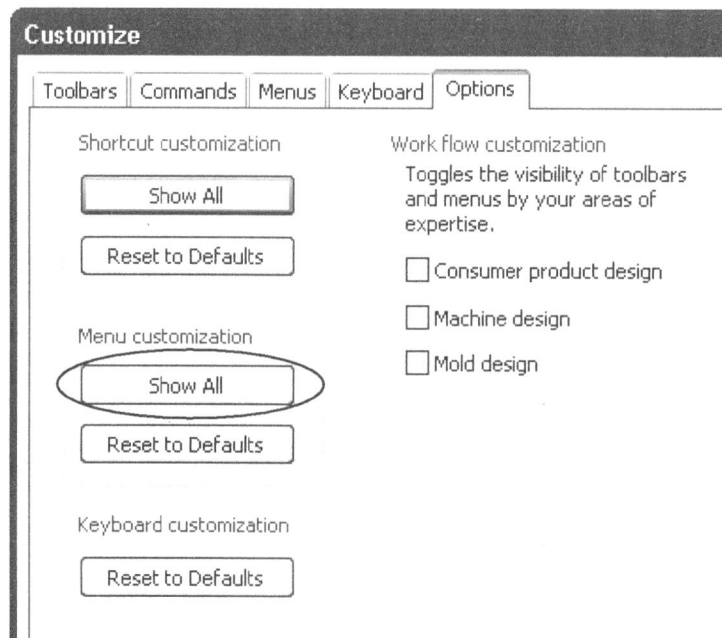

Now, all items are displayed in each menu. All of them are checked. The "Edit" menu is shown.

To reset the menus, pull down the "Tools" menu and pick **Customize** or right click over the toolbars and pick **Customize**.

In the **Customize** dialog box, click on the **Options** tab.

Under **Menu customization**, click the **Reset to Defaults** button and then click **OK**.

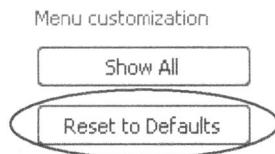

Menu customization

Show All

Reset to Defaults

All the menus are reset to the system default display. Note that the list of items in the "Edit" menu shown below is now much shorter.

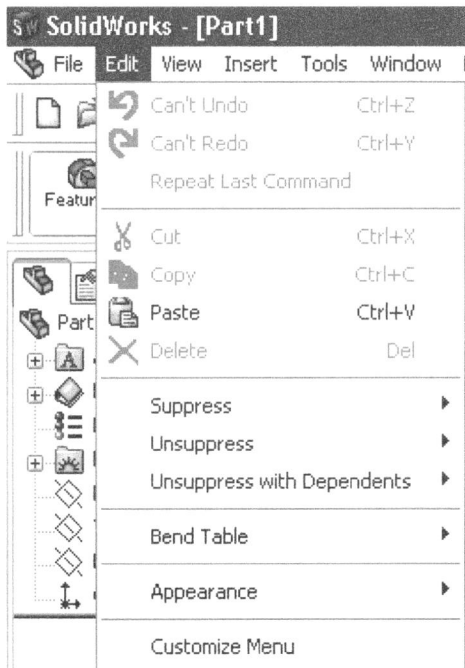

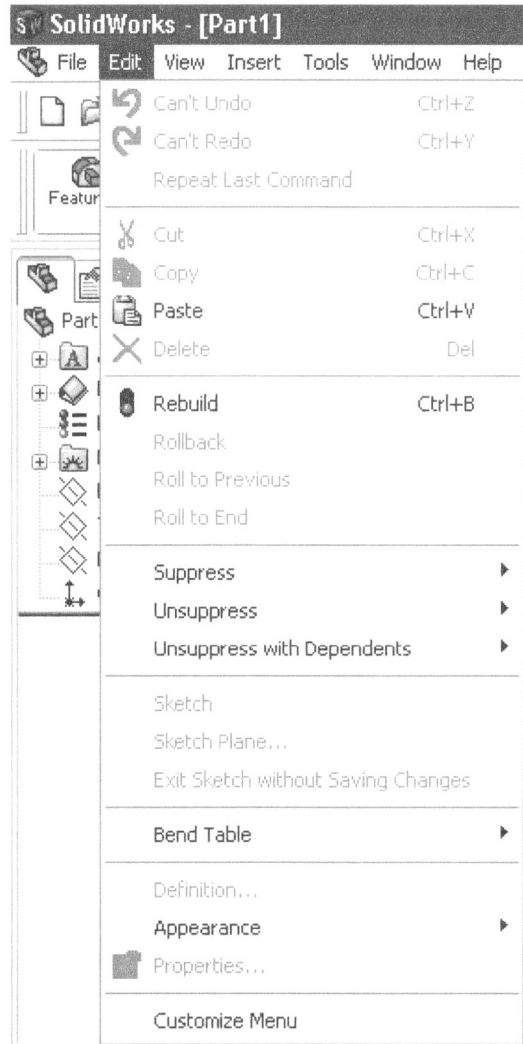

Advanced Menu Customization

A more advanced way of customizing the menus is to actually control what options are available in each of the pull down menus. Say that you often use a command that is buried in submenus. You can put it at the top level of a menu to get quicker access. Note that although you are able to do this, you have to manually undo any changes you make. **Reset to Defaults** does not do anything to the changes made on the **Menus** tab.

Pull down the "Tools" menu and pick **Customize** or right click over the toolbars and pick **Customize**.

In the **Customize** dialog box, click on the **Menus** tab.

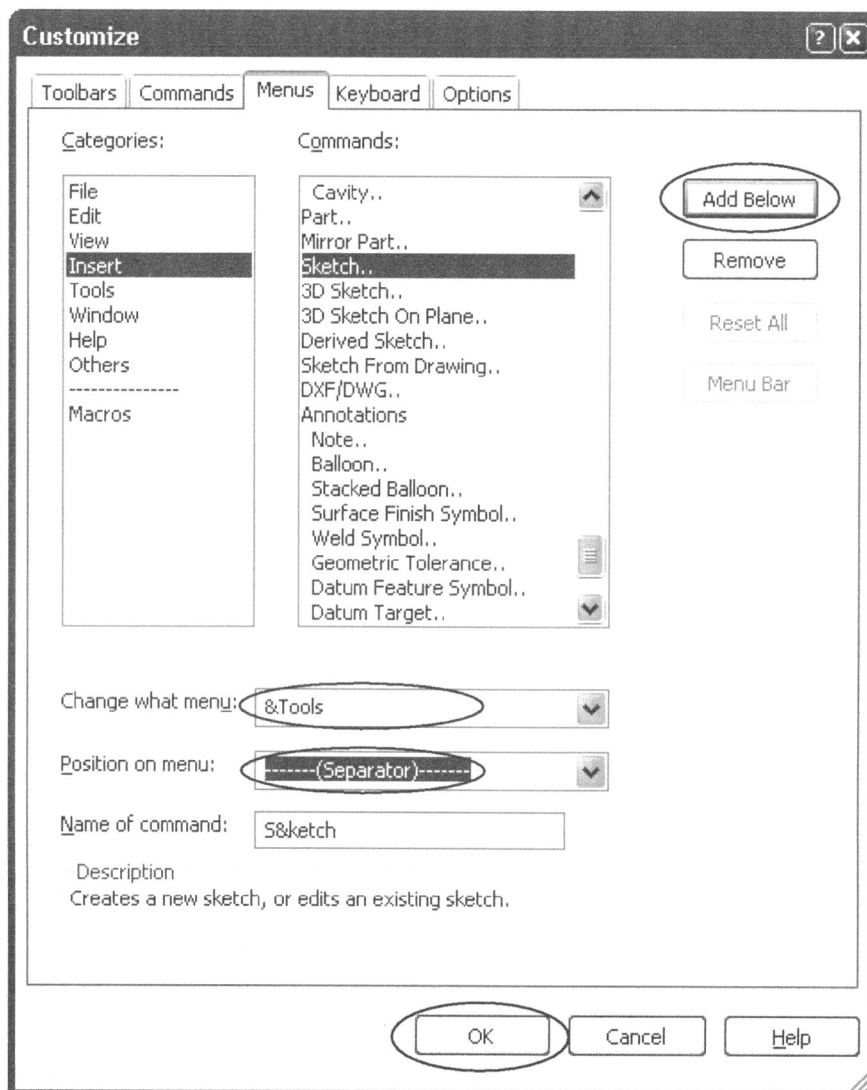

Here you can add any menu item into another menu or remove any item from a menu. You can also reorder the menus by adding and removing the items to create your own special order. Again, don't get too carried away. It is very frustrating trying to get things back the way they were.

The **Sketch** command is located in the "Insert" pull down menu already. Let's add it into the "Tools" pull down menu as well.

To do this, in the **Categories** box, pick **Insert**.

In the **Commands** box, scroll down to find and pick **Sketch..**

Pull down the **Change what menu** and pick **&Tools**.

Pull down the **Position on menu** and pick the **separator** just above the **S&ketch Entities** item.

Then click the **Add below** button followed by the **OK** button.

Now pull down the "Tools" menu. The **Sketch** command appears above the other sketch related commands.

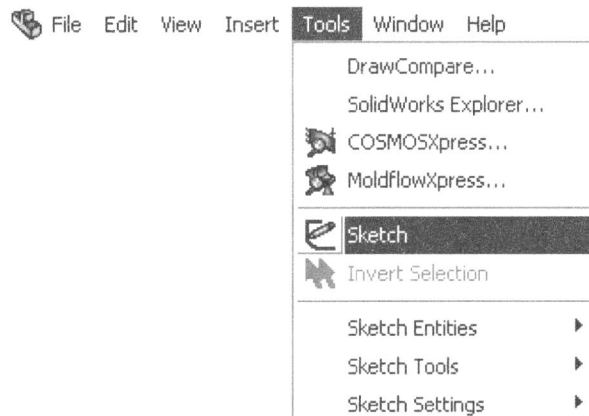

To remove an item from the menu, pull down the "Tools" menu and pick **Customize** or right click over the toolbars and pick **Customize**.

In the **Customize** dialog box, click on the **Menus** tab.

Pull down the **Change what menu** and pick **&Tools**.

Pull down the **Position on menu** and pick **Sketch**.

Then click the **Remove** button followed by the **OK** button.

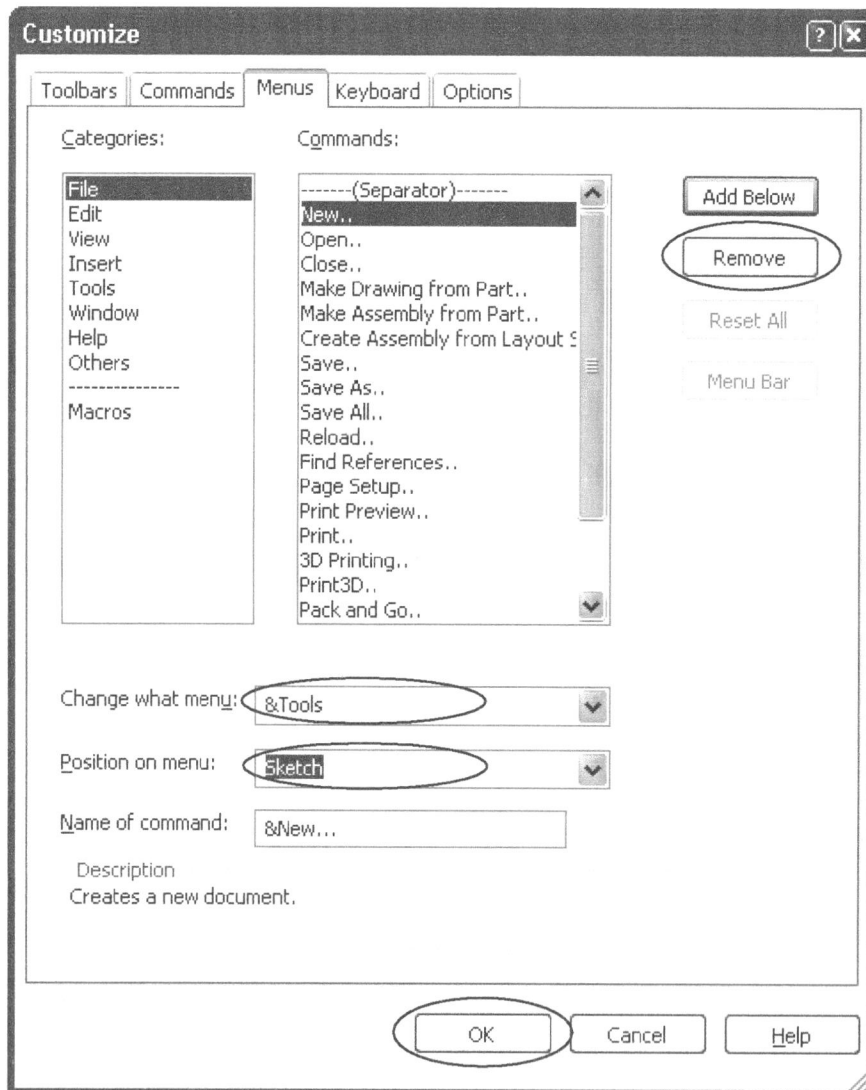

Remember, you have to manually undo any changes you make. Therefore, it is NOT recommended that you customize the menu commands any further unless you really know what you are doing. A better way to quickly access commands is through the use of custom toolbars and hot keys.

If you do change too much, you can use the Copy Settings Wizard to restore your settings. That's why we saved them before you started on this book.

Chapter 4

Another way to increase your efficiency using SolidWorks is to customize the toolbars. Toolbars allow you to create custom working environments where the functions you use regularly are just a mouse click away. Remember that too many toolbars and buttons can be confusing. Grouping your commonly used buttons and eliminating those that you rarely use will greatly enhance your performance.

You can customize the toolbars in SolidWorks by adding, removing, and moving buttons. Buttons can be added to a toolbar, which may be placed in any location. The buttons are easy to remember and easy to spot. The toolbars can be resized or moved, providing one-click access to commands. Almost any function is available through the use of toolbars. Toolbars can be visible when needed or hidden when not needed. SolidWorks remembers which toolbars are displayed and their location for each document type. You must, therefore, customize the toolbars for each document type. These customizations are remembered from session to session. So, if you make a toolbar visible in a part document, that toolbar is visible when you open a new part document, but not in an assembly or drawing. You cannot customize the name of a toolbar from the standard dialog box, but you are able to add any command to a toolbar.

In this chapter, you will customize a seldom used toolbar to include some of your favorite commands. This will then become your primary toolbar since it will contain all of your favorites. OK, what you are really doing here is only a sample to show what you can do. You can them use the technique to make it your favorite toolbar.

Create a New Part Document

If you do not have a part document open, click the **New** button in the "Standard" toolbar.

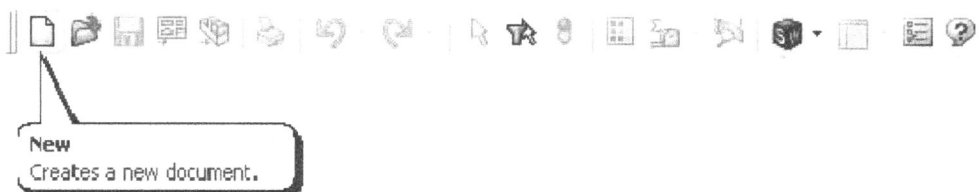

New
Creates a new document.

The **New SolidWorks Document** dialog box appears.

Click **Part** and then click **OK**. You may also just double click on **Part**.

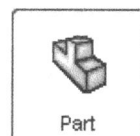

Part

Display a Toolbar

First, you will display the toolbar that you want to customize.

Pull down the "Tools" menu and pick **Customize**.

In the **Customize** dialog box, on the **Toolbars** tab, check **Explode Sketch** to display the "Explode Sketch" toolbar.

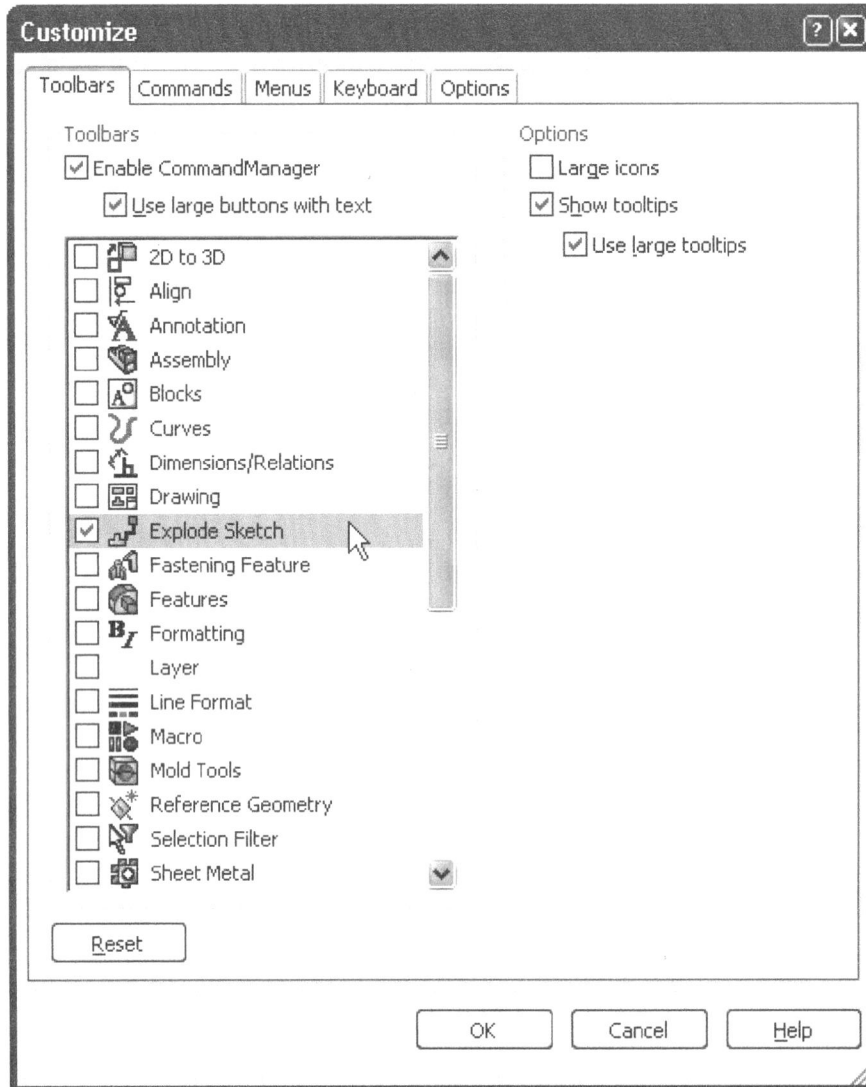

The "Explode Sketch" toolbar appears docked on the left side of the SolidWorks window. This is a small toolbar with only two buttons. So, you may need to look around for it on the screen. Once you find the "Explode Sketch" toolbar, undock it by placing the cursor over the two lines at the start of the toolbar. Remember that the cursor will change to a four-way arrow cursor. Hold down the left mouse button and drag the toolbar into the middle of the graphics area and release the mouse button.

Remove a Button from a Toolbar

When the **Customize** dialog box is open, you are in edit mode and can drag and drop buttons as you please. To remove a button from a toolbar, you simply drag the button off the toolbar while in edit mode.

Drag the **Jog Line** button outside the toolbar. The cursor will show a red X as shown, indicating that the button will be removed from the toolbar. Release the mouse button to remove the button from the toolbar. Note that there is no undo when removing buttons from a toolbar. To re-add a button you must find the command and re-add it manually, as shown in the next step.

Add a Button to a Toolbar

Click on the **Commands** tab in the **Customize** dialog box.

Under **Categories**, click on **Explode Sketch** to display the buttons from the "Explode Sketch" toolbar.

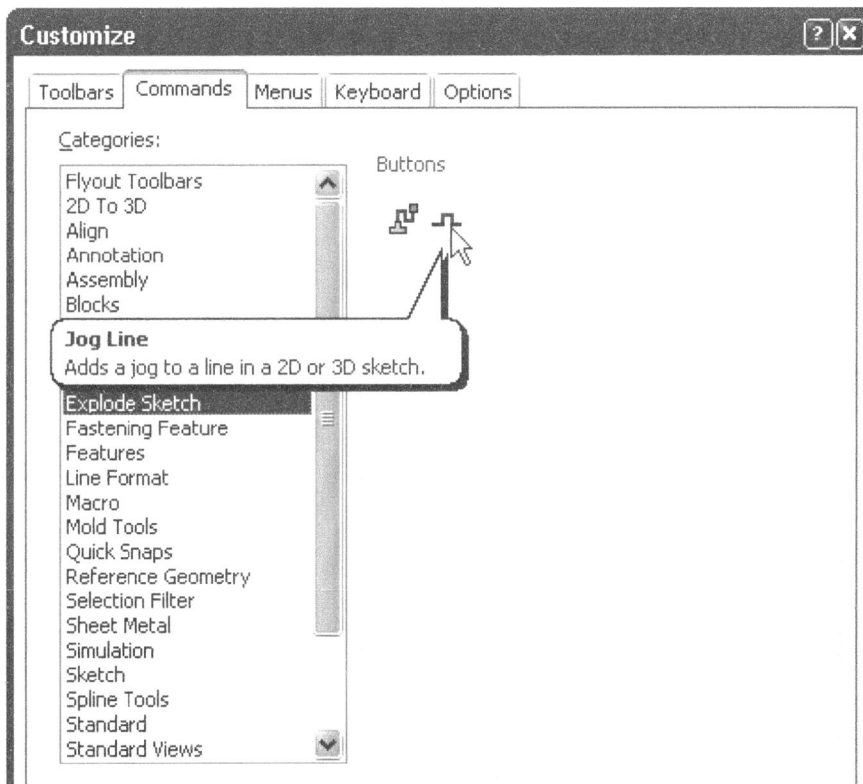

Under **Buttons**, drag the **Jog Line** button from the dialog box back onto the "Explode Sketch" toolbar, as shown above.

Create a Custom Toolbar

In this section, you will use an existing toolbar and modify it to your liking. Various commands throughout SolidWorks will be added to the custom toolbar. It is a great idea to place all the commands that you use frequently in one convenient location. Another advantage to creating a custom toolbar is that you can pick and choose commands from other toolbars. Once those commands are in your custom toolbar, the original toolbar does not have to be displayed, freeing up space to increase the size of the graphics area. Since the "Explode Sketch" toolbar only has two buttons, it is really easy to use for a custom toolbar.

On the "Explode Sketch" toolbar, drag the **Jog Line** button off the toolbar. The cursor will show a red X as shown, indicating that the button will be removed from the toolbar.

Then, drag the **Route Line** button off the toolbar. The cursor will show a red X as shown, indicating that the button will be removed from the toolbar.

The "Explode Sketch" toolbar should now be empty, as shown.

If you need to get at either of these commands, the **Route Line** command can be accessed by pulling down the "Tools" menu and picking **Sketch Entities**. In an open sketch with a line, or in a drawing with a sketched line, pull down the "Tools" menu and pick **Sketch Tools – Jog Line** to access the **Jog Line** command.

In the **Customize** dialog box, on the **Commands** tab, under **Categories**, click on **Standard**.

Under **Buttons**, drag the **New** button onto the "Explode Sketch" toolbar, as shown.

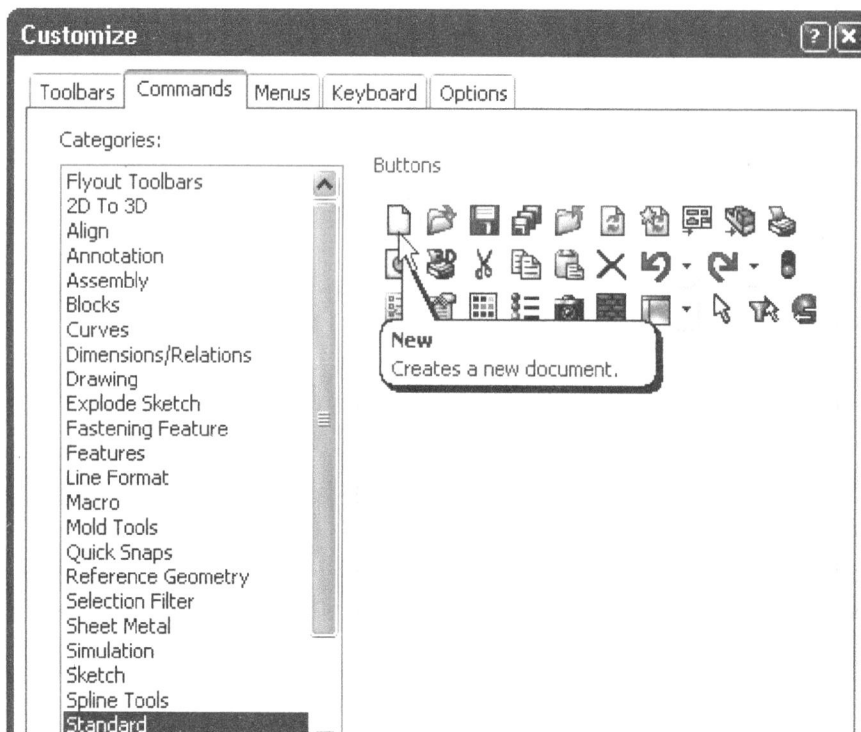

Under **Buttons**, drag the **Open** button onto the "Explode Sketch" toolbar, as shown.

In the **Customize** dialog box, on the **Commands** tab, under **Categories**, click on **Sketch**.

Under **Buttons**, drag the **Sketch** button onto the "Explode Sketch" toolbar, as shown.

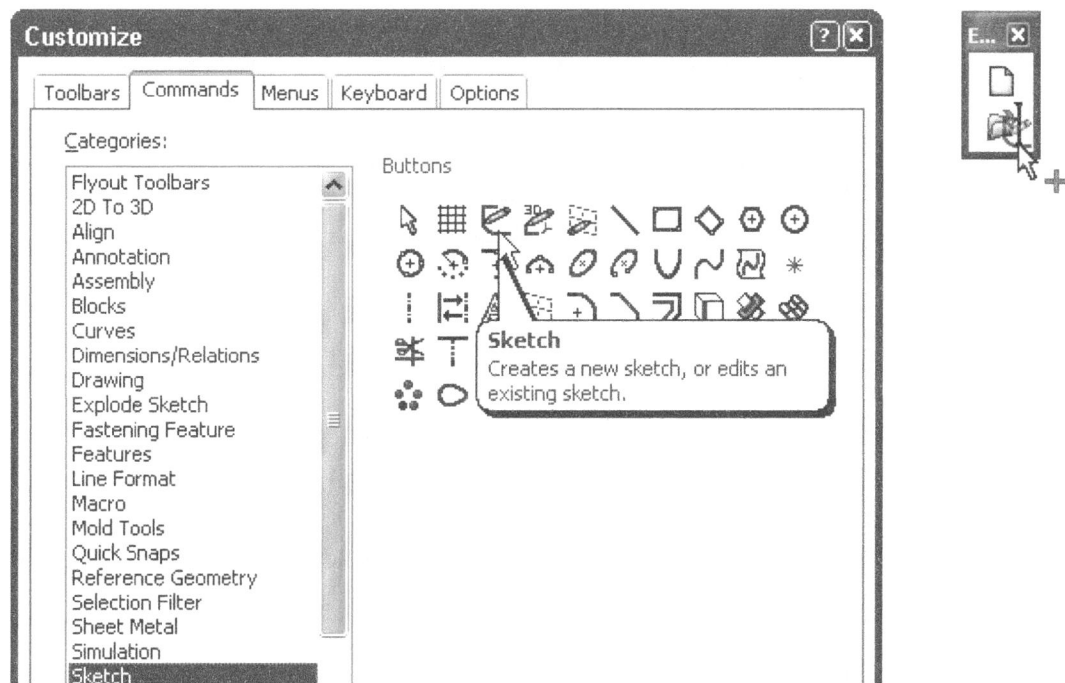

In the **Customize** dialog box, on the **Commands** tab, under **Categories**, click on **Features**.

Under **Buttons**, drag the **Extruded Boss/Base** button onto the "Explode Sketch" toolbar, as shown.

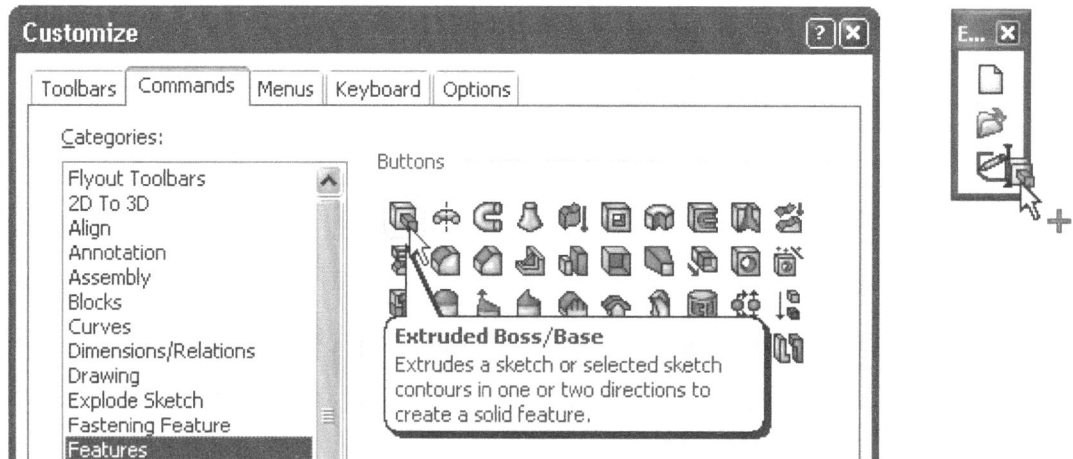

Under **Buttons**, drag the **Extruded Cut** button onto the "Explode Sketch" toolbar, as shown.

Under **Buttons**, drag the **Fillet** button onto the "Explode Sketch" toolbar, as shown.

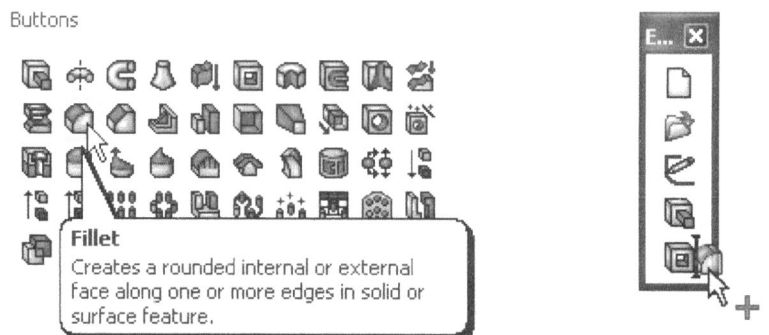

Under **Buttons**, drag the **Shell** button onto the "Explode Sketch" toolbar, as shown.

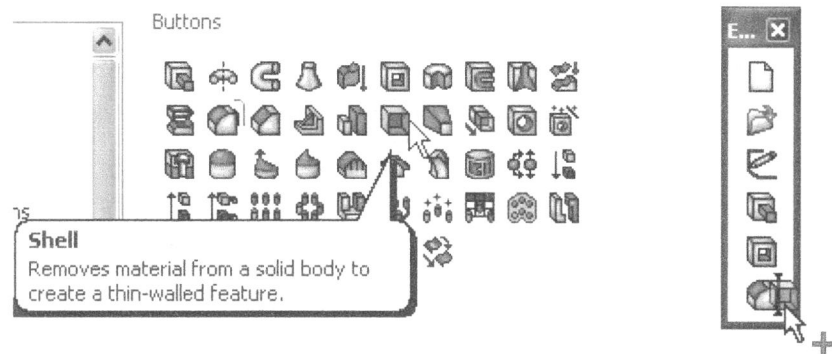

Shell
Removes material from a solid body to create a thin-walled feature.

Under **Buttons**, drag the **Rib** button onto the "Explode Sketch" toolbar, as shown.

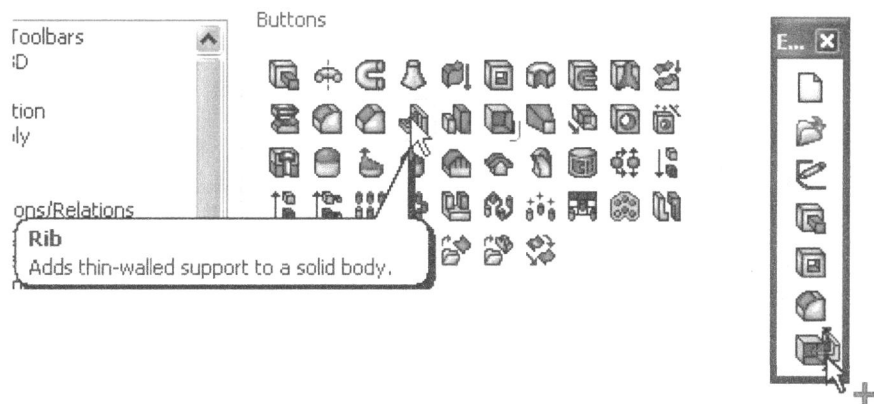

Rib
Adds thin-walled support to a solid body.

In the **Customize** dialog box, on the **Commands** tab, under **Categories**, click on **Reference Geometry**.

Under **Buttons**, drag the **Plane** button onto the "Explode Sketch" toolbar, as shown.

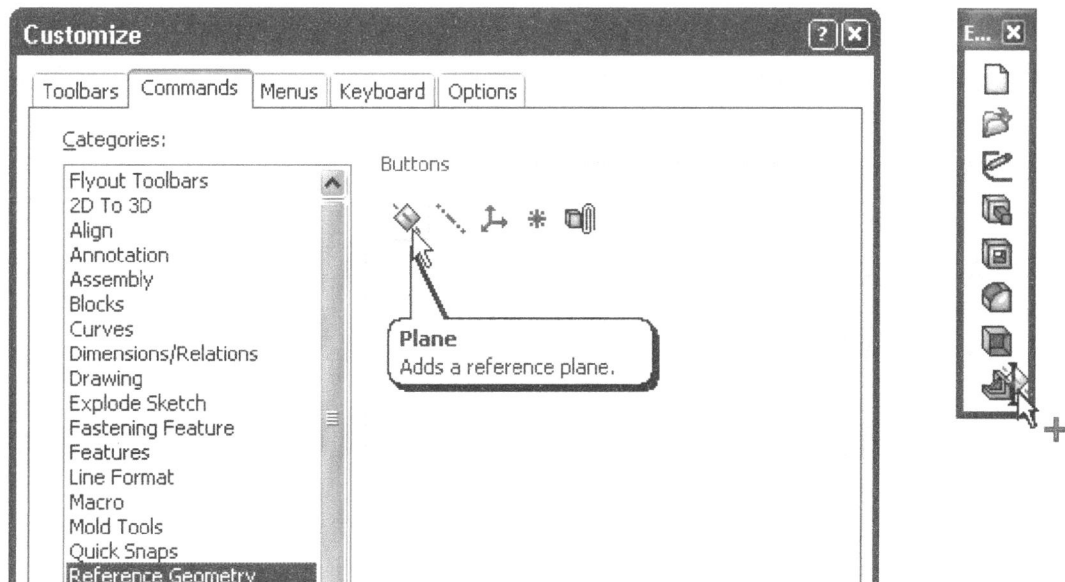

Plane
Adds a reference plane.

In the **Customize** dialog box, on the **Commands** tab, under **Categories**, click on **Flyout Toolbars**. The flyout toolbar buttons are just condensed toolbars. Each flyout toolbar button contains all the commands of the matching toolbar as a flyout menu. Clicking the button opens the flyout list of buttons.

Under **Buttons**, drag the **Surfaces** button onto the "Explode Sketch" toolbar, as shown.

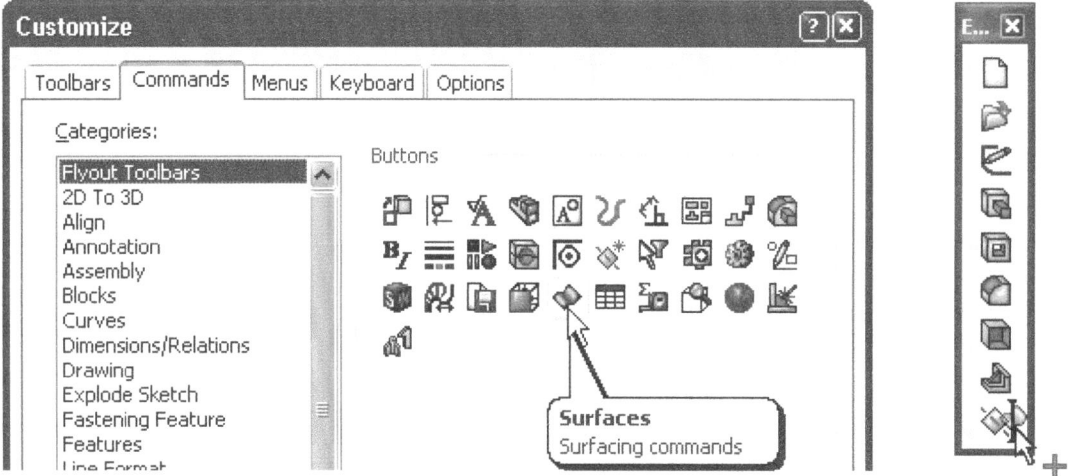

Under **Buttons**, drag the **Standard Views** button onto the "Explode Sketch" toolbar, as shown.

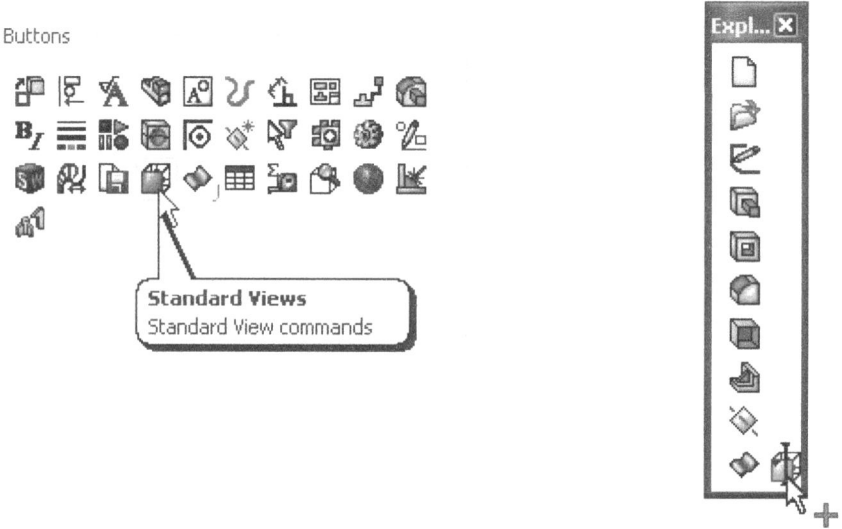

Finally, in the **Customize** dialog box, click **OK**. You now have a new toolbar customized just the way you wanted it.

Resize the Toolbar

To resize the "Explode Sketch" toolbar or change its orientation as shown below, click and drag one of the sides of the toolbar.

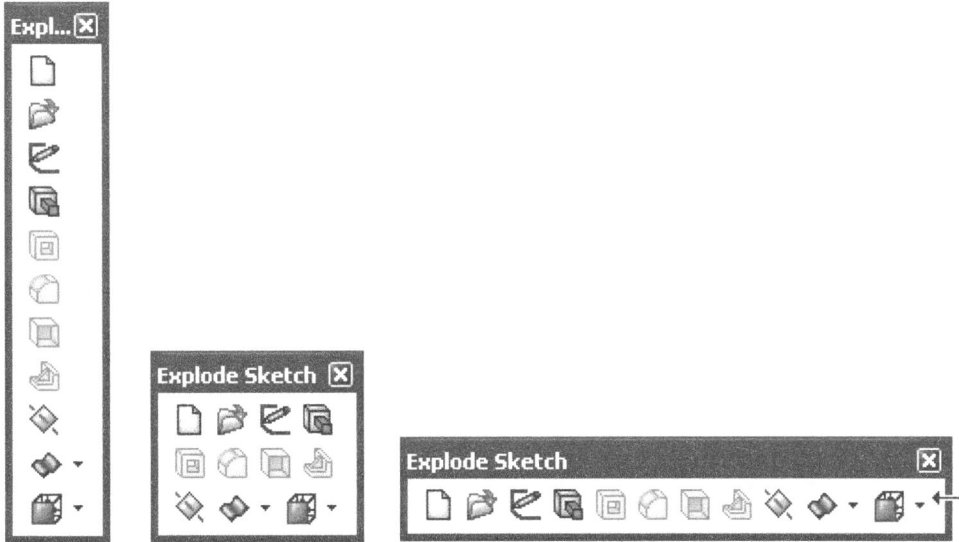

Quick Customization Hints

When the **Customize** dialog box is not open, you may still customize the toolbars. Simply hold down the **Alt** key and drag a button to move it. Drag and drop a button on a toolbar to add it to that toolbar

Add the **Hole Wizard** button from the CommandManager to your customized "Explode Sketch" toolbar.

To do this, hold down the **Alt** key and drag the **Hole Wizard** button from the CommandManager to where you want to place it on your customized "Explode Sketch" toolbar as shown, and release the mouse button. Be careful! Make sure that the cursor feedback does not show a red X. This will remove the button from the toolbar.

The only way to cancel the action is to place the cursor over the original button location, in this case the **Hole Wizard** button, and release the mouse button.

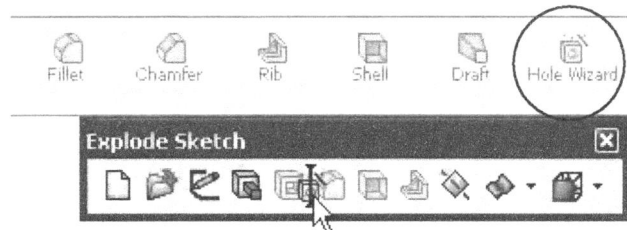

Notice that using the **Alt** key method moves the button; it does not copy the button. The button is no longer on the original toolbar. Now, hold down the **Alt** key and drag the **Hole Wizard** button on your customized "Explode Sketch" toolbar back onto the CommandManager. Place it next to the **Draft** button as shown, and release the mouse button.

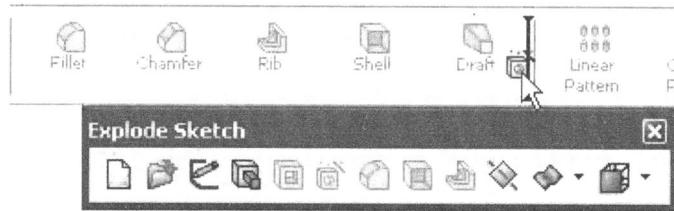

If you drop a non-copied button anywhere else, the button is removed from the toolbar. There is no undo. If you accidentally remove a button from a toolbar, you must go back to the **Customize** dialog box, find the button, and add it back to where it was removed, as shown below.

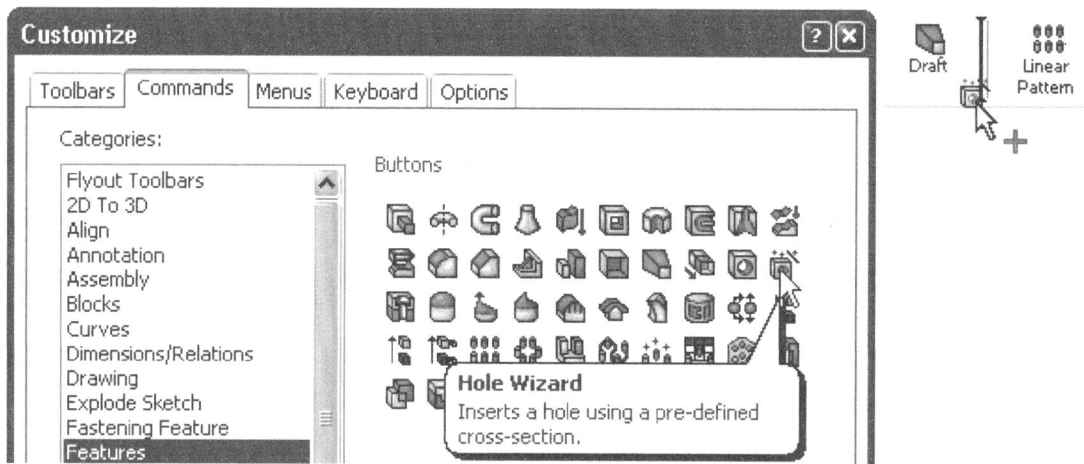

Reorder a Button on a Toolbar

To reorder the buttons on the toolbar, hold down the **Alt** key and drag the **Cut-Extrude** button to the right of the **Shell** button, as shown.

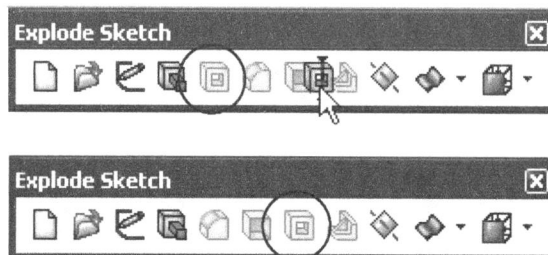

Drag the "Explode Sketch" toolbar below the CommandManager to dock the toolbar horizontally.

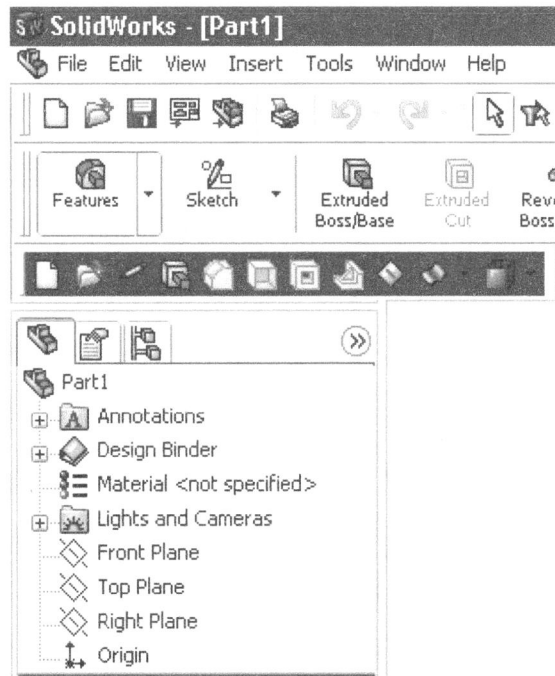

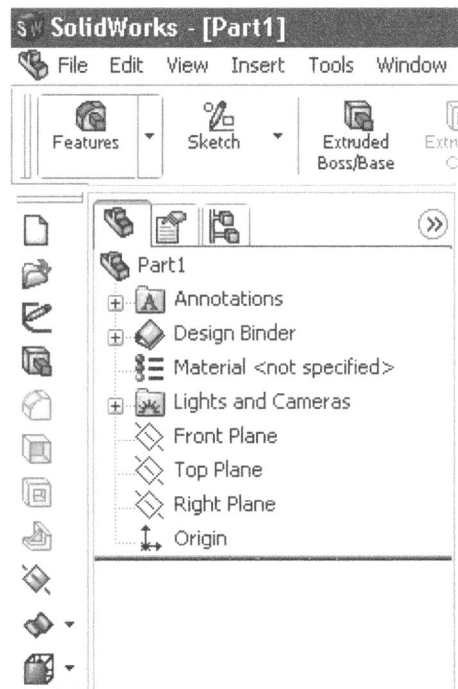

Drag the "Explode Sketch" toolbar to the left side of the screen to dock the toolbar vertically. Slide the toolbar up to the CommandManager.

Right click on your new customized, docked toolbar and click **Standard** to uncheck, or hide, the toolbar.

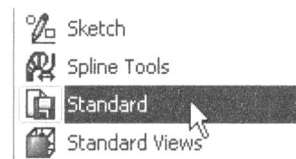

Then, right click again on your new customized, docked toolbar and click **View** to uncheck, or hide, the toolbar.

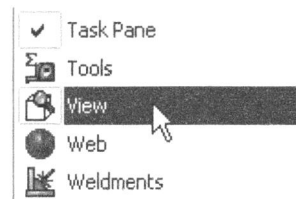

Remember, the location and display of the toolbars is only for the currently active document type, meaning that the look and feel is only changed for part documents since you are in a part file.

CommandManager

The CommandManager can be used to access toolbar buttons in a central location and save space for the graphics area. By default, it contains toolbars based on the document type. When you click a button in the control area on the left end of the CommandManager, the right end updates to show that toolbar. So, if you click **Sketch** in the control area, the "Sketch" toolbar appears to the right in the CommandManager. This helps reduce the number of toolbars you need to display.

To get started, make sure that the CommandManager is enabled by pulling down the "Tools" menu and picking **Customize**.

In the **Customize** dialog box, on the **Toolbars** tab, make sure the **Enable CommandManager** check box is checked. If it is not, click the check box to check it, and then click **OK**.

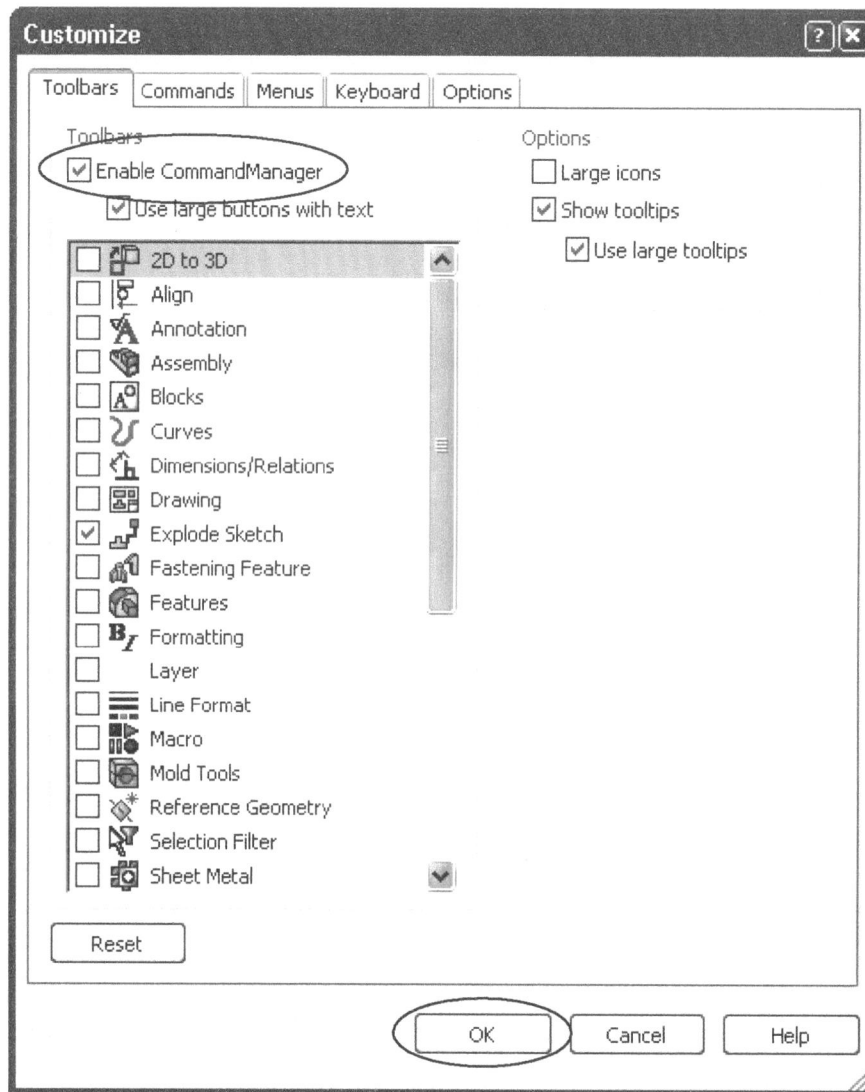

There is a faster way to enable or disable the CommandManager. Simply right click over a toolbar, not on the CommandManager. A menu will appear to control the display of the toolbars. At the top of the list is **CommandManager**. A check next to this choice indicates the CommandManager is active. If it is not checked, pick **CommandManager** to display the CommandManager.

Add a Toolbar to the CommandManager

There are a couple of ways to customize the CommandManager. One way is to move the cursor over the CommandManager and click the right mouse button. Then pick **Customize Command Manager**.

A menu appears listing the available toolbars which could be shown in the CommandManager. Pick **Standard Views** from the list to check this toolbar, and then click in the graphics area, or press **Enter**. From this list, you do not have control over the order that the toolbars will appear in the CommandManager.

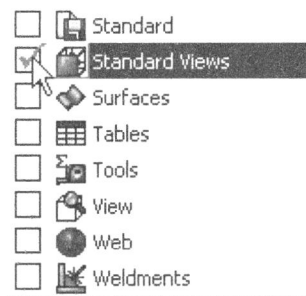

In the CommandManager, click **Standard Views**. The "Standard Views" toolbar is now available. You can also right-click on the CommandManager and select which toolbar to activate.

Adding too many toolbars to the CommandManager is not a good thing. Since each toolbar added increases the control area of the CommandManager, reducing the available space to display the toolbar buttons on the right. You could solve this by undocking the CommandManager and resizing it, but that would take away from the graphics area.

Customize the Toolbar Order in the CommandManager

You can easily move the toolbar's position and add toolbars onto the CommandManager by using the same technique as you would to move buttons around on any other toolbar. Also, the CommandManager itself can be moved to any location, just like a regular toolbar.

Place the cursor at the beginning of the customized "Explode Sketch" toolbar. Drag the toolbar onto the beginning (left side) of the CommandManager, as shown. A plus sign appears on the cursor to indicate that you will add the toolbar to the CommandManager.

Now you do have control of the order your toolbars in the CommandManager. hold down the **Alt** key and drag the buttons to reorder them as you see fit.

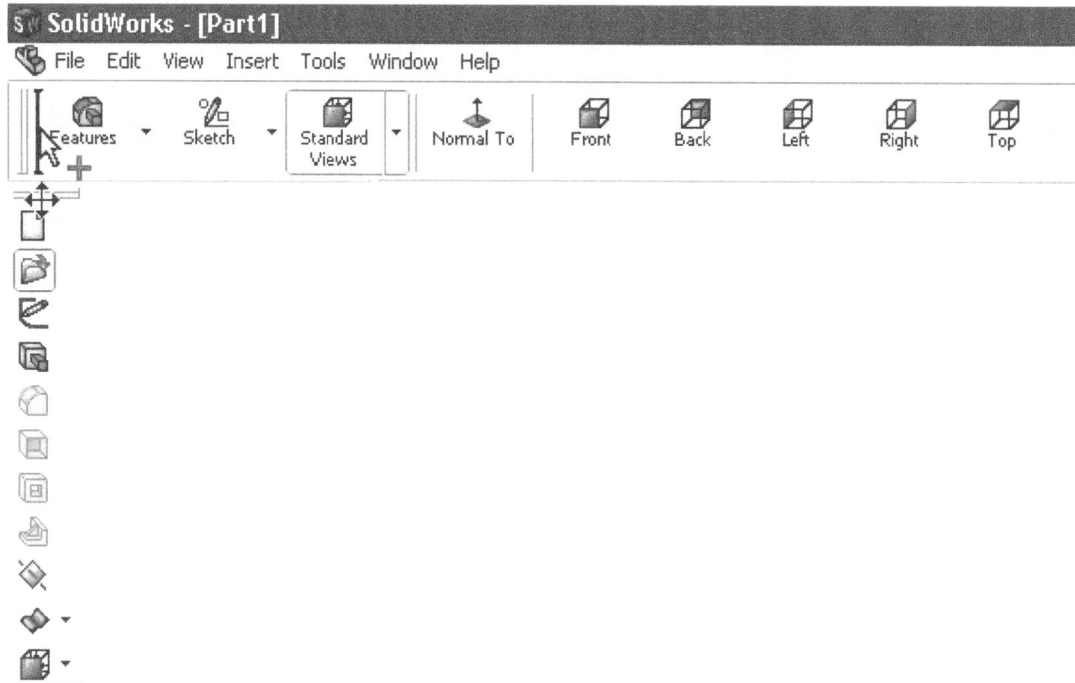

On the CommandManager, click **Explode Sketch**. The "Explode Sketch" toolbar is now available.

Notice that the "Explode Sketch" toolbar was not copied, but it was moved onto the CommandManager. The display of the "Explode Sketch" toolbar was turned off. You can easily turn it back on by right clicking over the menu bar and checking **Explode Sketch** from the menu. For now, leave the display of the toolbar turned off.

Use Large Buttons with Text

To customize the CommandManager further, right click over the CommandManager and pick **Use Large Buttons with Text** to uncheck it.

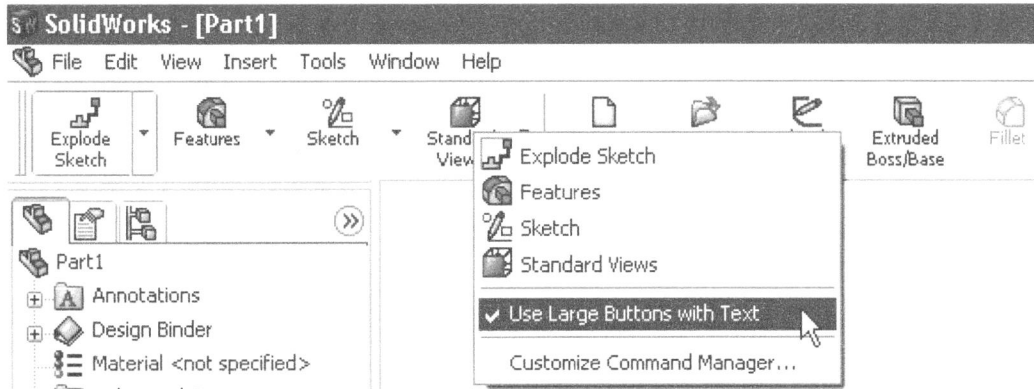

The CommandManager now is shown with only buttons and no text.

This option is also available by right clicking on the menu bar and picking **Customize**, or by pulling down the "Tools" menu and picking **Customize**.

On the **Toolbars** tab, under **Toolbars**, check **Use large buttons with text** to return the CommandManager to the original display.

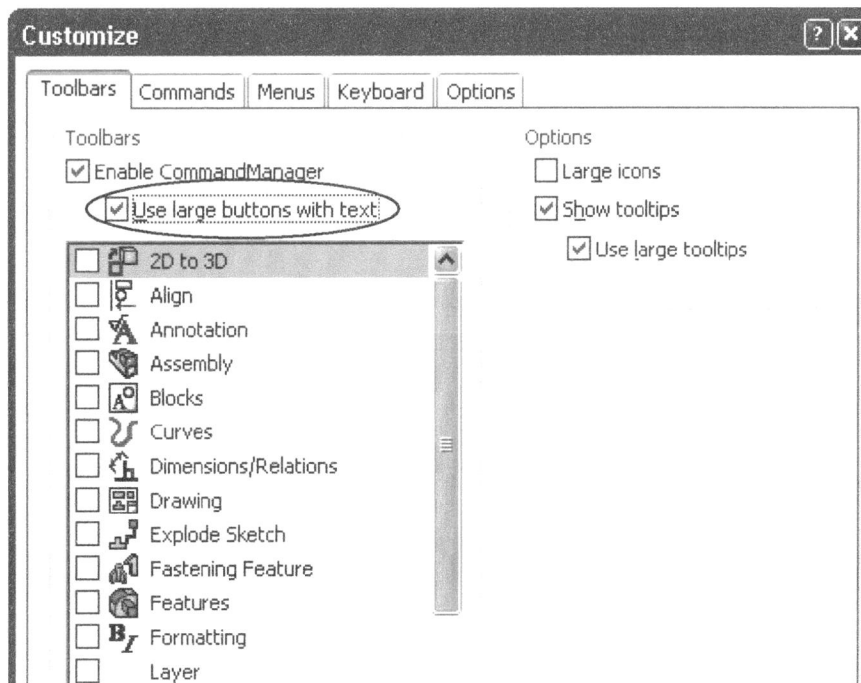

Unchecking **Enable CommandManager** clears the CommandManager, and the toolbars in the CommandManager appear in the SolidWorks interface.

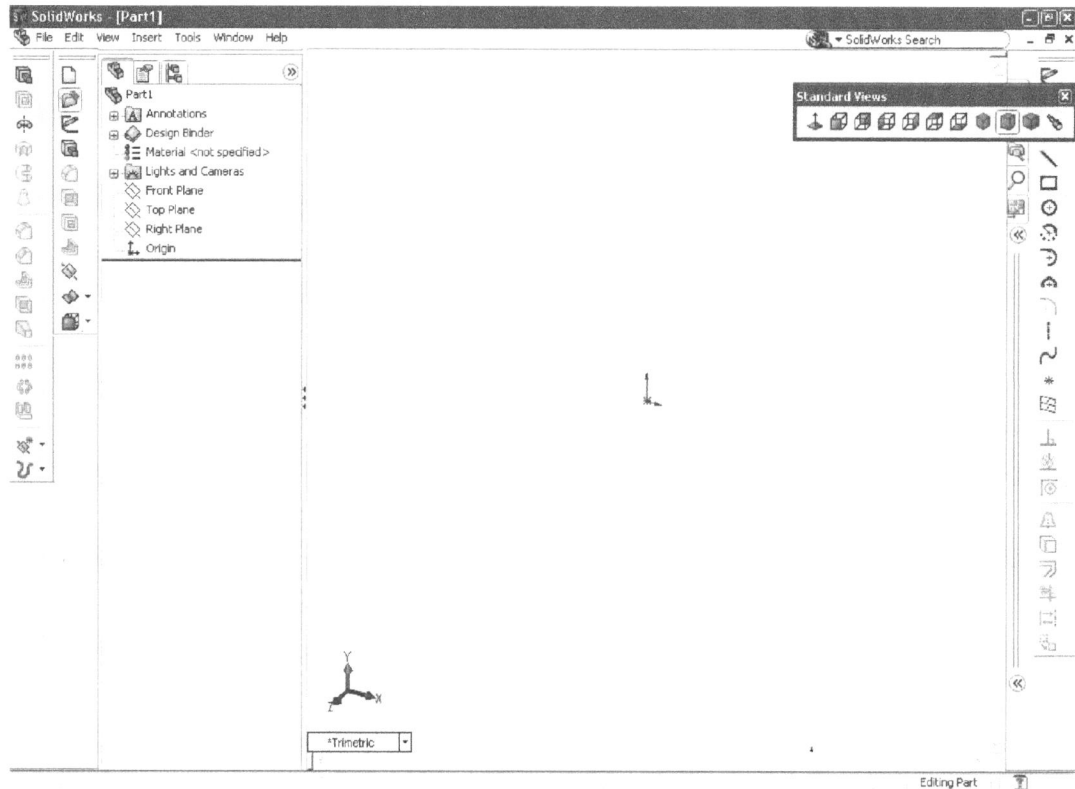

Keep **Enable CommandManager** checked and click **OK**.

Remove Toolbars from the CommandManager

Remove the "Standard Views" toolbar from the CommandManager by right clicking on the CommandManager and picking **Customize Command Manager**. Click **Standard Views** in the list to uncheck it, and then click in the graphics area, or press **Enter**.

Add Commands to a CommandManager Toolbar

Add a command to the "Sketch" toolbar by clicking **Sketch** in the CommandManager.

Then, right click on the menu bar and pick **Customize**, or pull down the "Tools" menu and pick **Customize**.

In the **Customize** dialog box, click on the **Commands** tab.

Under **Categories**, click on **Sketch**.

Drag the **Polygon** button onto the CommandManager to the right of the **Rectangle** button.

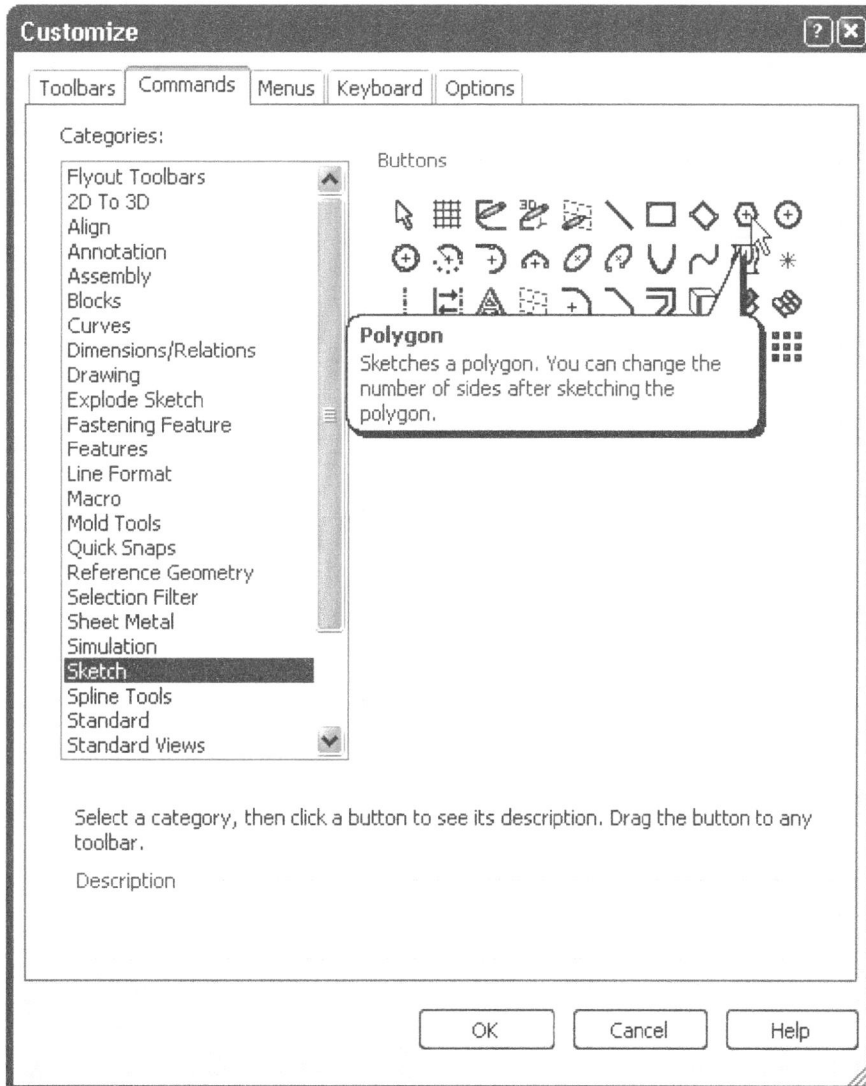

Click **OK** to close the **Customize** dialog box.

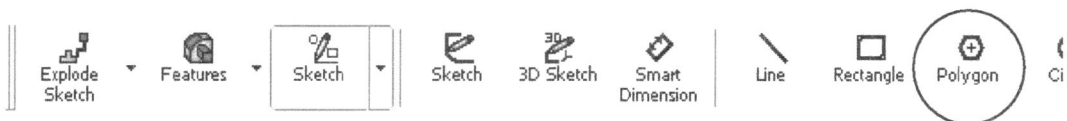

Customizing a toolbar in the CommandManager is really the same as customizing it as a toolbar.

Right click on a toolbar and click **Sketch** to display the "Sketch" toolbar. The **Polygon** button is in this toolbar because you added it through the Command Manager.

Right click on a toolbar and click **Sketch** to hide the "Sketch" toolbar.

The CommandManager itself is not a toolbar, but rather a collection of toolbars providing you access to multiple toolbars without filling your screen with more buttons and toolbars than you can handle. This is a powerful tool which when customized to your style of work, will save you time and effort throughout the day. And remember, the CommandManager is different in an assembly and a drawing, so you will want to take advantage of it in those documents as well.

Chapter 5

In this chapter you will be focusing on increasing your productivity with keyboard shortcuts, pressing two or three keys simultaneously to perform an action. Keyboard shortcuts are intended to speed up your use of SolidWorks. How much time this will actually save depends on you. Most people can save about a second using a shortcut rather than locating, moving the cursor to, and clicking on a button. They can save a fraction more if the command has to be located in a pull down menu. If you can save about one second hundreds of times a day, and keep more of your attention on the design rather than the movements to perform commands, then a lot of your time will be saved.

There are a few reasons why you should use keyboard shortcuts. For example, do you use the standard Windows shortcuts for cut, copy, paste (Ctrl+X, Ctrl+C, Ctrl+V) instead of looking for the buttons with the cursor or searching through the pull down menus? Since you can use your one hand to perform the shortcut, you don't have to take your other hand off of the mouse. This will speed up your work. Keyboard shortcuts are quick and easy to use. Once you use the shortcuts enough, you can pretty much do them without even thinking about it, just like someone who is a real fast typist doesn't have to look at the keyboard. They are obviously faster than anyone staring at the keyboard for each and every letter they type. Also, there are shortcuts that you use all the time, and then there are others that you don't remember because you only used it once. A simple rule is if a shortcut saves time, it should be used. If it doesn't, then don't bother. Remember that you must memorize the shortcut or reference a cheat sheet to be effective. However, once you become familiar with your keyboard shortcuts, you'll easily forget how time consuming it was trying to find your favorite commands in the menus.

While many users already use keyboard shortcuts, many miss the optional ability to control the display of toolbars with keyboard shortcuts. The clever user can quickly create a powerful interface where the use of keyboard shortcuts can quickly toggle the display of custom toolbars.

Use Keyboard Shortcuts

SolidWorks includes pre-assigned shortcut keys for commonly used commands. These keyboard shortcuts can save you time and effort to execute frequently used commands. Try a few of them out when creating this simple part.

If a part document is not open, create a new **Part** file by holding down the **Ctrl** key and pressing the **N** key (**Ctrl+N**).

In the **New SolidWorks Document** dialog box, double click on **Part**.

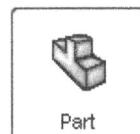

Part

Create the base feature by clicking the **Explode Sketch** button in the control area of the CommandManager. Then, click the **Extruded Boss/Base** button from the toolbar, or pull down the "Insert" menu and pick **Boss/Base – Extrude**.

Extruded Boss/Base
Extrudes a sketch or selected sketch contours in one or two directions to create a solid feature.

Select the **Top** plane when prompted to select a plane on which to sketch the feature cross-section.

Front Plane

Top Plane

Top Plane

Create a rectangle with the origin inside the rectangle using the **Rectangle** button in the CommandManager, or pull down the "Tools" menu and pick **Sketch Entities – Rectangle**.

Create a construction line diagonally across the rectangle by clicking the **Centerline** button in the CommandManager, or pull down the "Tools" menu and pick **Sketch Entities – Centerline**.

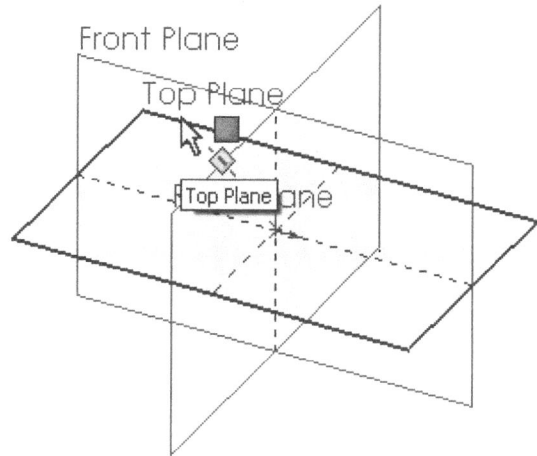

Click the top left corner and then the bottom right corner to create the centerline.

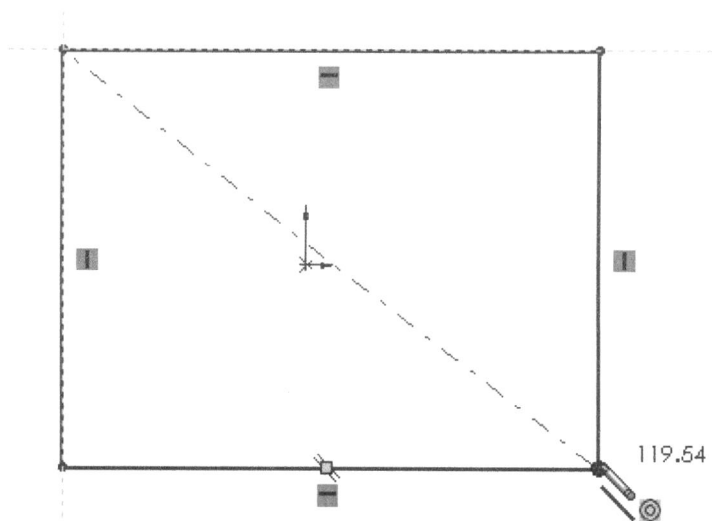

119.54

Press the **Escape** key to deselect the **Centerline** tool.

Select the diagonal line. Then, hold down the **Ctrl** key and select the origin.

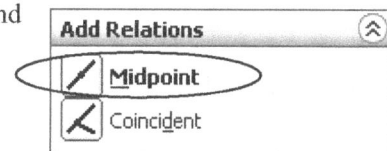

Add Relations

Midpoint

Coincident

In the **Properties** PropertyManager, under **Add Relations**, click the **Midpoint** button.

Click the **Smart Dimension** button in the CommandManager, or pull down the "Tools" menu and pick **Dimensions – Smart.**

Add a '**10**' vertical dimension to the left vertical line and a '**24**' horizontal dimension between the left and right side of the sketch. You may need to press the **F** key to **Zoom to Fit**.

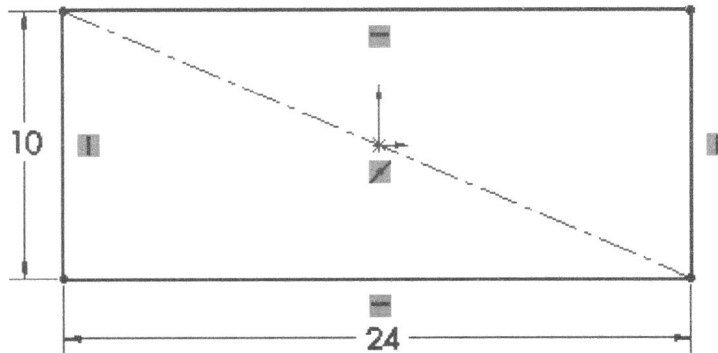

Exit the sketch by clicking the **Exit Sketch** button in the CommandManager or in the upper right corner of the graphics area.

In the **Extrude** PropertyManager, under **Direction 1**, set the **Depth** to '**5**'.

Click the green check mark button at the top of the **Extrude** PropertyManager to accept the settings and create the feature.

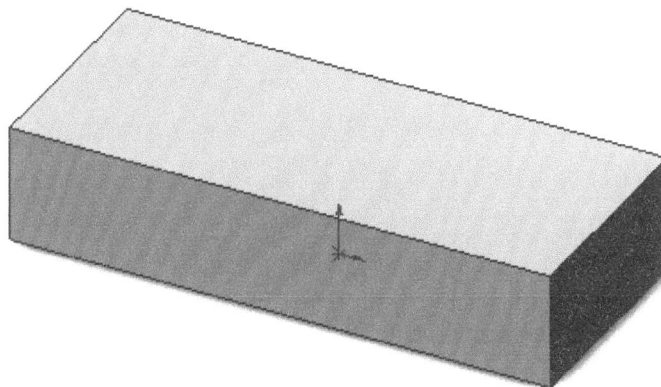

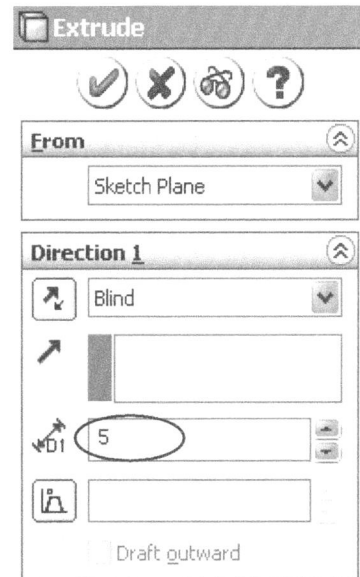

Create a Shortcut

Before you create a keyboard shortcut for a function, you have to think about what makes a shortcut worth your time. First, you want to make a shortcut something that you can easily remember. A good suggestion is to associate the first letter of the function to the shortcut, like **Ctrl+P** for **Print**, **Ctrl+O** for **Open**, or **Z** for **Zoom**. Second, the best keyboard shortcuts can be done with one hand, allowing you to keep your one hand on the mouse and your other hand close to the commonly used **Ctrl**, **Tab**, and **Esc** keys. You can easily create your own shortcuts to most of the commands in SolidWorks. However, it is best to create shortcuts only for the most commonly used functions.

Pull down the "Tools" menu and pick **Customize**, or right click on the menu bar and pick **Customize**.

In the **Customize** dialog box, click on the **Keyboard** tab.

Scroll down to the **Category Tools** and find **Customize**, and then click in the **Shortcut(s)** box.

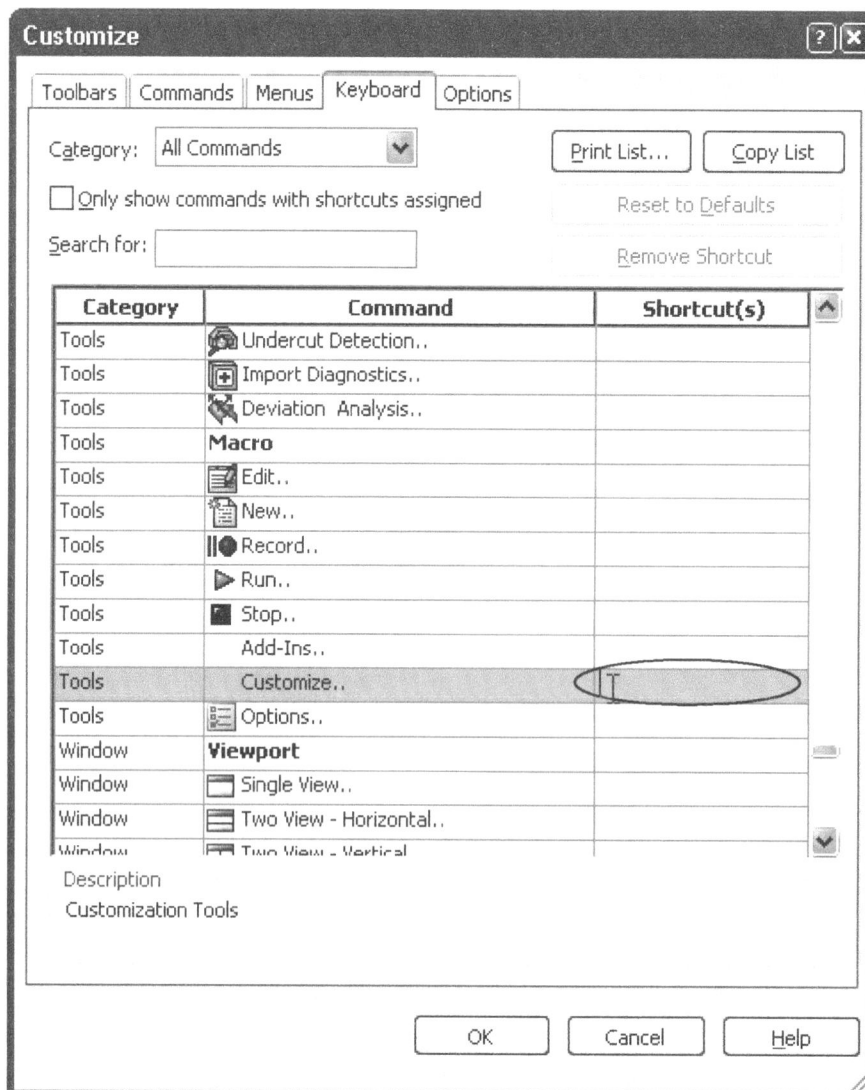

Category	Command	Shortcut(s)
Tools	Undercut Detection..	
Tools	Import Diagnostics..	
Tools	Deviation Analysis..	
Tools	**Macro**	
Tools	Edit..	
Tools	New..	
Tools	Record..	
Tools	Run..	
Tools	Stop..	
Tools	Add-Ins..	
Tools	Customize..	
Tools	Options..	
Window	**Viewport**	
Window	Single View..	
Window	Two View - Horizontal..	
Window	Two View - Vertical	

Description
Customization Tools

Press the **C** key. Note that if a shortcut key is already assigned, a dialog box appears. If you choose to use the shortcut for the new command, it is removed from the old command. It is not a good practice to overwrite SolidWorks or Windows default shortcuts.

In the **SolidWorks** dialog box, click **No**.

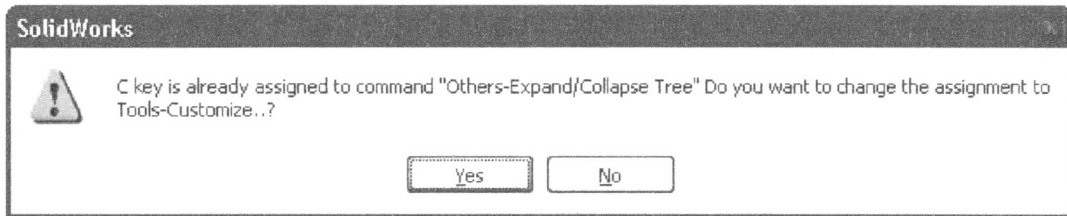

SolidWorks

⚠️ C key is already assigned to command "Others-Expand/Collapse Tree" Do you want to change the assignment to Tools-Customize..?

[Yes]   [No]

Click in the **Shortcut(s)** box again, and hold down the **Alt** key and press **C**. **Alt+C** appears as the shortcut. Note that letters are shown in uppercase. The **Caps Lock** key is not used and does not replace the **Shift** key.

Category	Command	Shortcut(s)
Tools	Undercut Detection..	
Tools	Import Diagnostics..	
Tools	Deviation Analysis..	
Tools	**Macro**	
Tools	Edit..	
Tools	New..	
Tools	Record..	
Tools	Run..	
Tools	Stop..	
Tools	Add-Ins..	
Tools	Customize..	Alt+C
Tools	Options..	
Window	**Viewport**	
Window	Single View..	
Window	Two View - Horizontal..	
Window	Two View - Vertical	

Press the **Backspace** key or click the **Remove Shortcut** button to remove the shortcut. Try to stay away from using the **Alt** key in a shortcut to avoid any conflict with the pull down menus. (Windows allows you to press the **Alt** key and then the corresponding letter key to access the pull down menu.)

Hold down the **Ctrl** key and the **Shift** key, and then, press the **C** key. **Ctrl+Shift+C** is displayed.

Note that multiple shortcuts can be assigned to a command.

Category	Command	Shortcut(s)
Tools	Undercut Detection..	
Tools	Import Diagnostics..	
Tools	Deviation Analysis..	
Tools	**Macro**	
Tools	Edit..	
Tools	New..	
Tools	Record..	
Tools	Run..	
Tools	Stop..	
Tools	Add-Ins..	
Tools	Customize..	Ctrl+Shift+C
Tools	Options..	
Window	**Viewport**	
Window	Single View..	
Window	Two View - Horizontal..	
Window	Two View - Vertical	

In the **Customize** dialog box, click **OK**.

Search for a Command to Customize

Press **Ctrl+Shift+C** to open the **Customize** dialog box.

In the **Customize** dialog box, click on the **Keyboard** tab.

Click in the **Search for** box and type '**zoom**' to search all the commands containing "zoom."

Then, click in the **Shortcut(s)** box for **Zoom to Area** and press the **W** key. Feel free to create shortcuts that are a single letter. I used the **W** key because I think of this command as a window.

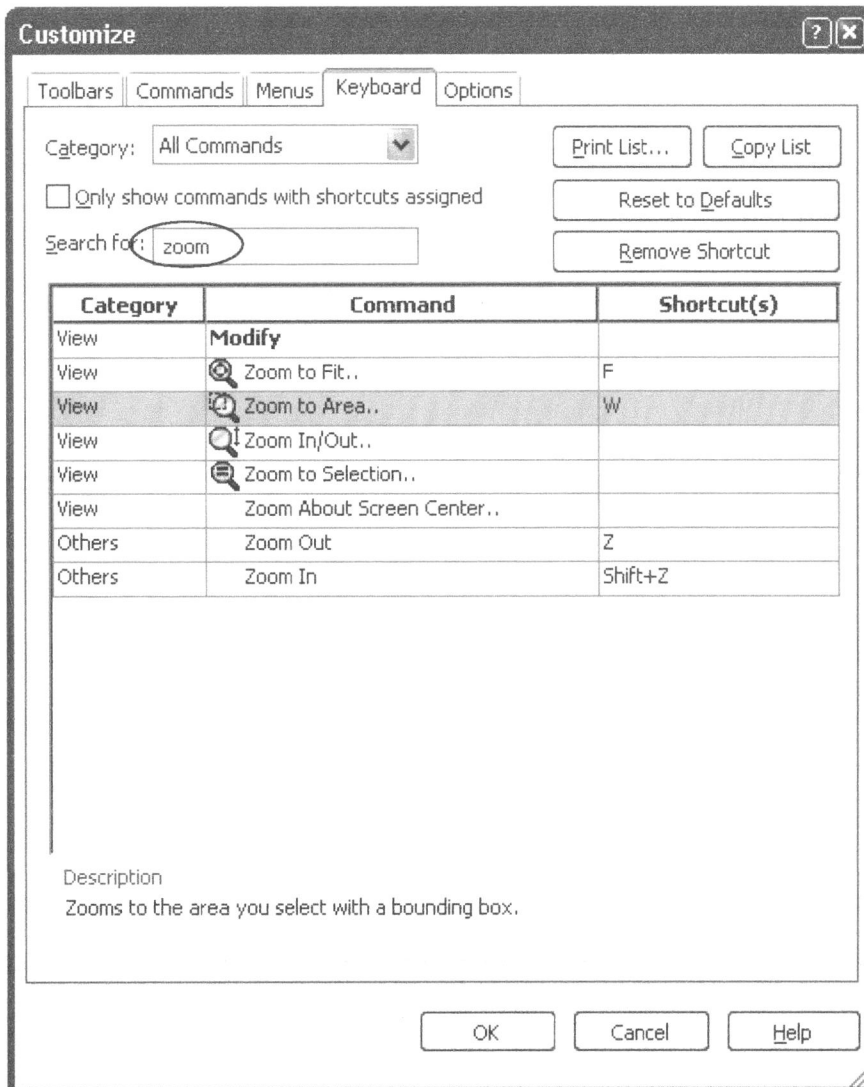

Customize

Toolbars | Commands | Menus | **Keyboard** | Options

Category: All Commands   Print List...   Copy List

☐ Only show commands with shortcuts assigned   Reset to Defaults

Search for: zoom   Remove Shortcut

Category	Command	Shortcut(s)
View	**Modify**	
View	🔍 Zoom to Fit..	F
View	🔍 Zoom to Area..	W
View	🔍 Zoom In/Out..	
View	🔍 Zoom to Selection..	
View	Zoom About Screen Center..	
Others	Zoom Out	Z
Others	Zoom In	Shift+Z

Description
Zooms to the area you select with a bounding box.

OK   Cancel   Help

Sort Customize Commands by Category

Delete **zoom** from the **Search for** box by deleting the text, and then, pull down the **Category** menu and pick **View**.

Scroll down until you find the "Explode Sketch" toolbar in the command list.

Then, click in the **Shortcut(s)** box for **Explode Sketch** and press the **1** key on the numeric keypad, with Number Lock turned on. Note that the **1** key on the alphabetic part of the keyboard doesn't work, only the numeric keypad on keyboards that have a separate number pad.

Customize

Toolbars | Commands | Menus | **Keyboard** | Options

Category: ⟨View⟩ ▼   [Print List...] [Copy List]

☐ Only show commands with shortcuts assigned

Search for: [_____]   [Reset to Defaults]

[Remove Shortcut]

Category	· Command	Shortcut(s)
View	**Toolbars**	
View	CommandManager..	
View	2D to 3D..	
View	Align..	
View	Annotation..	
View	Assembly..	
View	Blocks..	
View	Curves..	
View	Dimensions/Relations..	
View	Drawing..	
View	eDrawings 2007..	
View	Explode Sketch..	Num 1
View	Fastening Feature..	
View	Features..	
View	Formatting..	
View	Layer	

Description
Explode Sketch Toolbar

[OK] [Cancel] [Help]

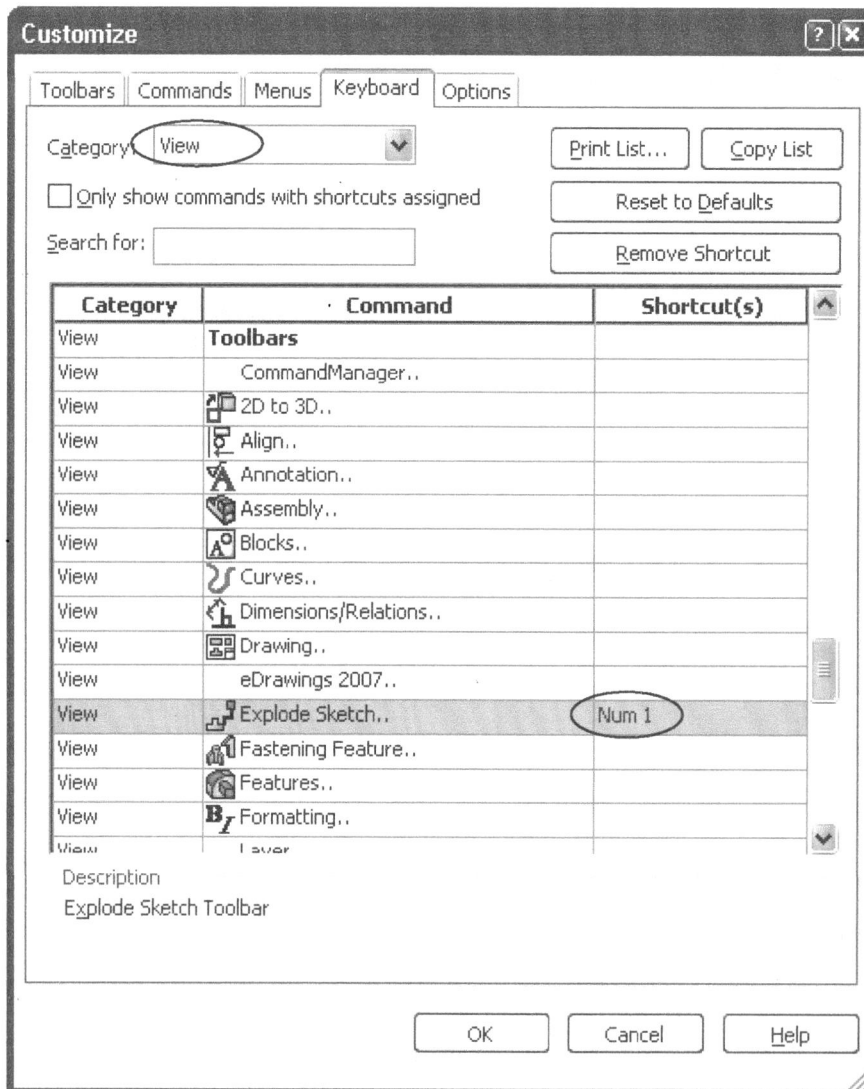

In the **Customize** dialog box, click **OK**.

Now, if you press the **1** key on the numeric keypad, the toolbar toggles on if it is not displayed, and off if it is displayed. This is a great way to gain drawing space but still have your tools close by. With a press of a key you can quickly display a set of tools on a customized toolbar or just a standard toolbar. After you are done using the toolbar, just press the key again to turn off the display of the toolbar.

Display Commands with Shortcuts

If you want to quickly see all your keyboard shortcuts, open the **Customize** dialog box by pressing **Ctrl+Shift+C**.

On the **Keyboard** tab, check **Only show commands with shortcuts assigned** to display the commands within the selected category (in this case, All Commands) that have shortcuts assigned.

You also have the option to print or copy the list.

Click the **Print List** button to open the **Print Setup** dialog box and print the selected list with columns for **Category**, **Command**, and **Shortcut(s)**.

Click the **Copy List** button to copy the selected list to the clipboard so you can paste (**Ctrl+V**) it into a document in Word, Excel, or another program of your choice.

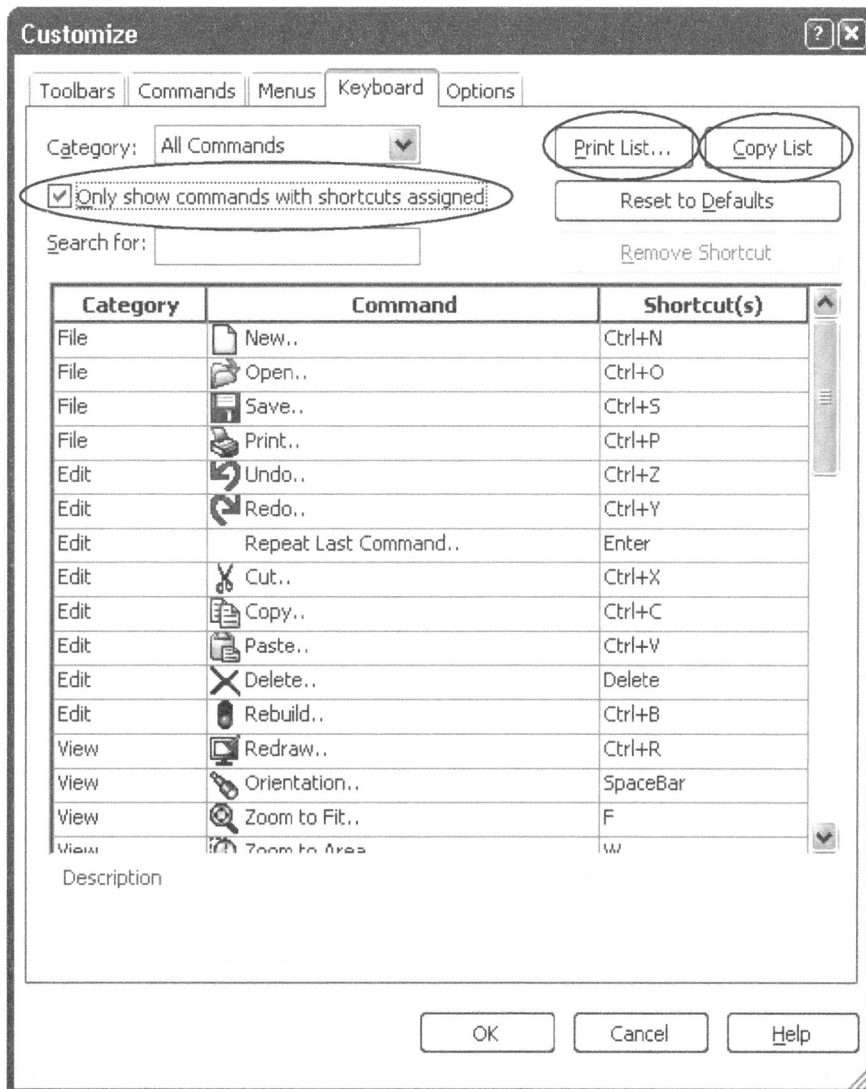

To abandon the current changes, click **Cancel**.

Save the File

Pull down the "File" menu and pick **Save**. Use '**block**' for the name of the file. You will use this part in the next chapter.

Reset All Shortcuts

To reset all shortcuts to the system defaults, open the **Customize** dialog box by pressing **Ctrl+Shift+C**.

Note that resetting to the defaults will remove all the shortcuts that you just added. DO NOT do this step right now. If you want to reset the shortcuts for any reason, you can click the **Reset to Defaults** button on the **Keyboard** tab. However, you will use some of the shortcuts that you just created throughout this book. So, DO NOT reset to the defaults.

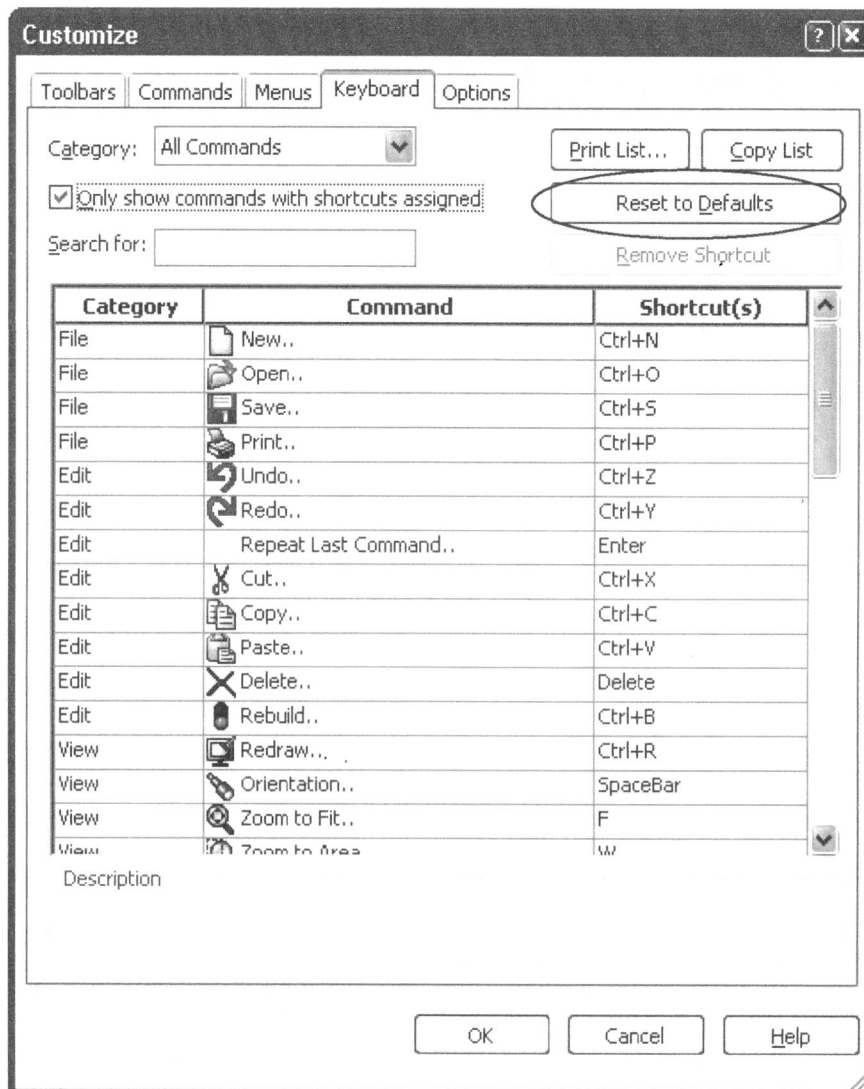

Chapter 6

Macros

Macros are a great way to reduce the number of steps that you do on a regular basis. You can save a lot of mouse clicks by recording a sequence of actions and replaying it over and over again as many times as you wish. Not only can you record the actions for playback, you can add a button to a toolbar to call the macro and assign a keyboard shortcut to it. Now you are really gaining speed!

Macros tend to fall into three areas. Some add new functionality that didn't exist before, while others simply enhance an existing feature. Finally, there are others designed to reduce steps.

Macros are limited in their ability to interact with you. They cannot pop up dialog boxes or other controls. They can prompt you with an InputBox() function or display a message with the MsgBox() function. However they are recorded as Visual Basic commands and a good programmer can write his own code around them to do pretty much anything he wants. But let' snot get carried away here.

As an example of a macro, you may need to display the annotations either for reference or to modify the dimensions. Using macros, you can record the steps (actions) to do this. Then, create custom buttons which will display and hide the annotations. Finally, keyboard shortcuts are assigned for quick and easy implementation of the macros.

Record a Macro

A macro allows you to record your mouse clicks, menu choices, and keystrokes for play back later, without having to have any programming knowledge.

If the **block** part document from the previous chapter is not open, open it by pressing **Ctrl+O**. In the **Open** dialog box, browse to where you saved the file and double click on **block**.

To record a macro, display the "Macro" toolbar by pulling down the "Tools" menu and picking **Customize**.

In the **Customize** dialog box, on the **Toolbars** tab, check **Macro**.

Another, faster, way to do this is to right click on the menu bar or a toolbar and pick **Macro**.

In the **Customize** dialog box, click **OK**.

Locate the "Macro" toolbar and drag it next to the CommandManager to dock it as shown.

In the "Macro" toolbar, click the **Record\Pause Macro** button, or pull down the "Tools" menu and pick **Macro – Record**.

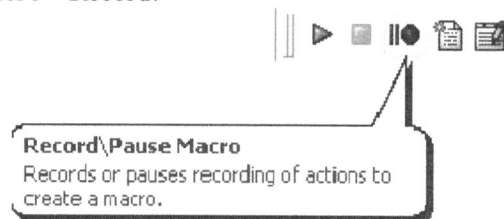

Record\Pause Macro
Records or pauses recording of actions to
create a macro.

Pull down the "Tools" menu and pick **Options**.

Click on the **Document Properties** tab.

Under **Detailing**, click on **Annotations Display**.

Check **Feature dimensions** and **Display annotations**, and then click **OK**.

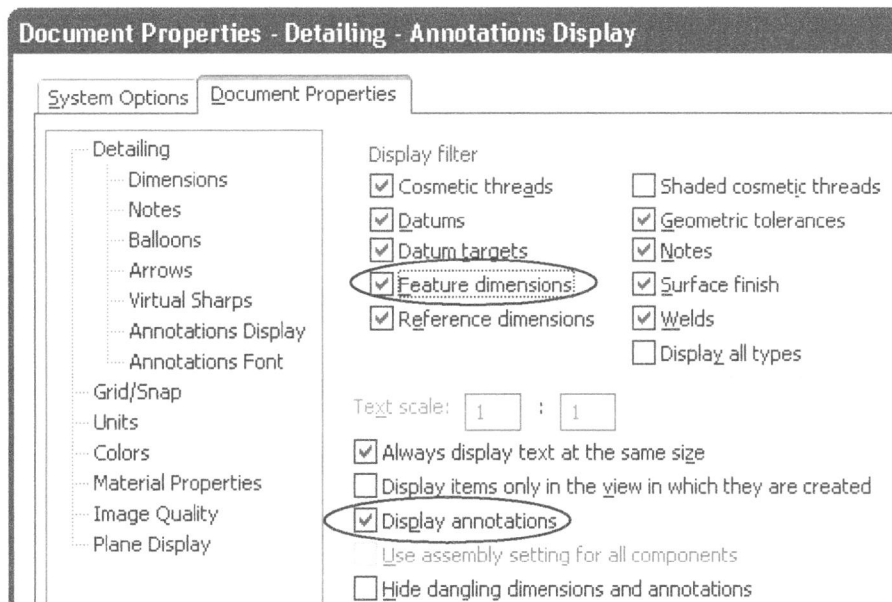

When you are done, click the **Stop Macro** button on the "Macro" toolbar, or pull down the "Tools" menu and pick **Macro – Stop**.

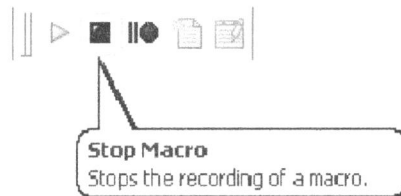

In the **Save As** dialog box, browse to the SolidWorks installation directory and click the **Create New Folder** button.

When the new folder appears, type '**Macros**' and press the **Enter** key. Before you assign a macro to a shortcut key or to a menu item, you must create a macro folder named **Macros** with an '**s**'.

Double click on the new **Macros** folder.

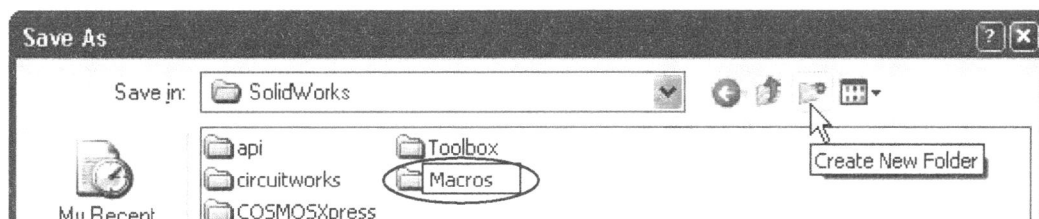

Click in the **File name** box and type '**Display Annotations**'. Then click **Save**. Note that macros are saved with the .swp extension.

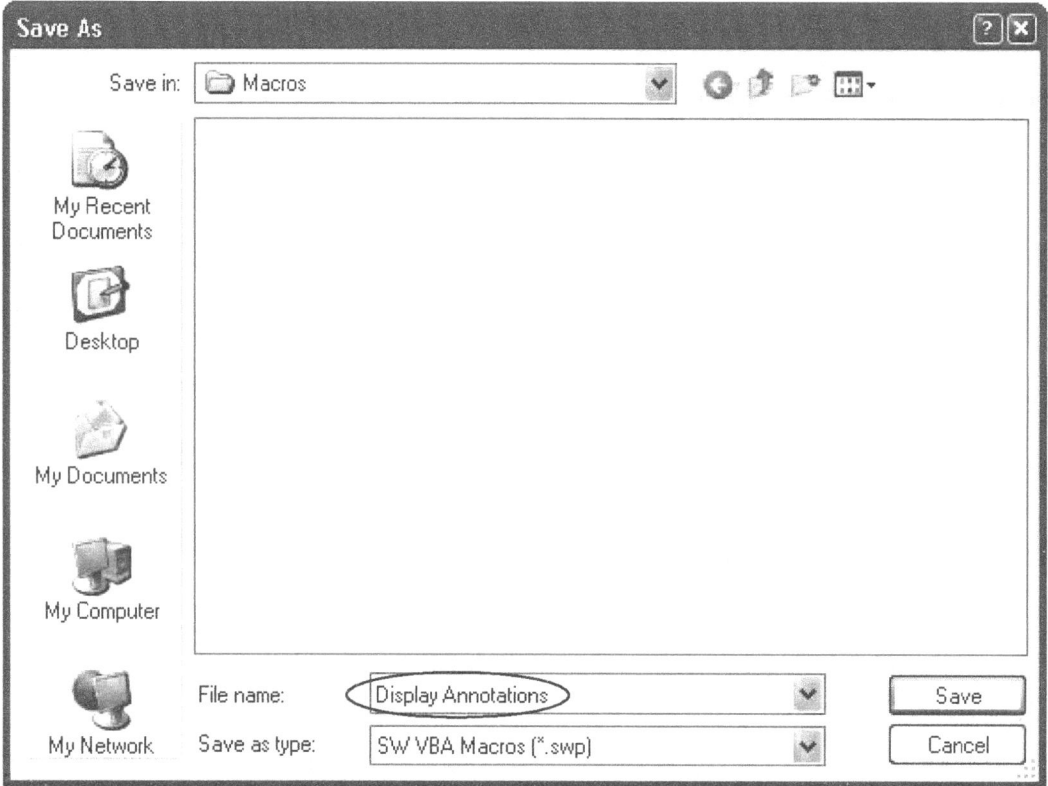

The dimensions of the Block part now appear on the screen. You can drag the dimensions to change their position and make them more visible, if you wish.

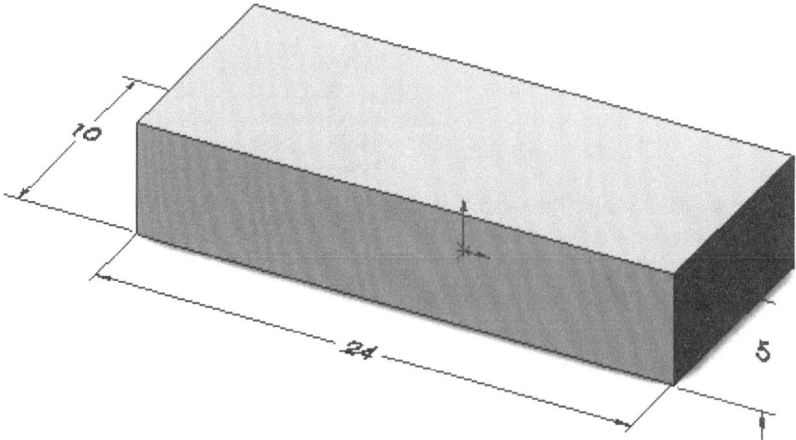

Record Another Macro

II● In the "Macro" toolbar, click the **Record\Pause Macro** button, or pull down the "Tools" menu and pick **Macro – Record**.

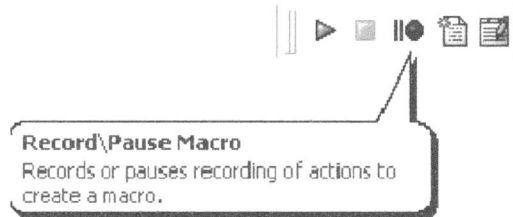

> **Record\Pause Macro**
> Records or pauses recording of actions to create a macro.

Pull down the "Tools" menu and pick **Options**.

Click on the **Document Properties** tab.

Under **Detailing**, click on **Annotations Display**.

Uncheck **Feature dimensions** and **Display annotations**, and then click **OK**.

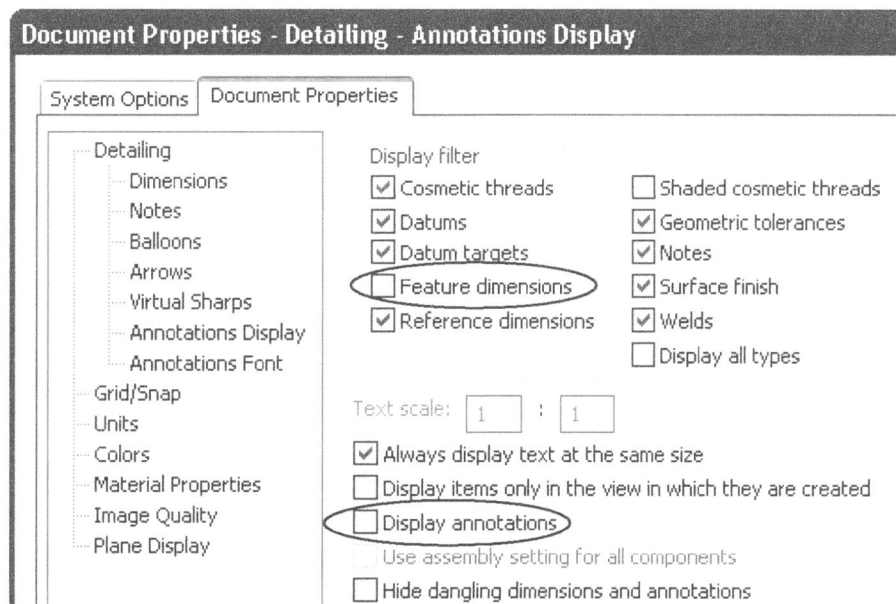

■ Click the **Stop Macro** button on the "Macro" toolbar, or pull down the "Tools" menu and pick **Macro – Stop**.

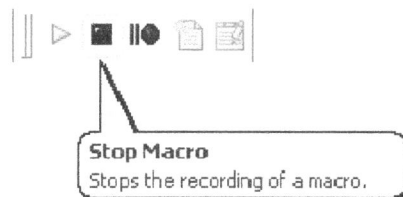

> **Stop Macro**
> Stops the recording of a macro.

In the **Save As** dialog box, click in the **File name** box and type '**Hide Annotations**'. Then click **Save**.

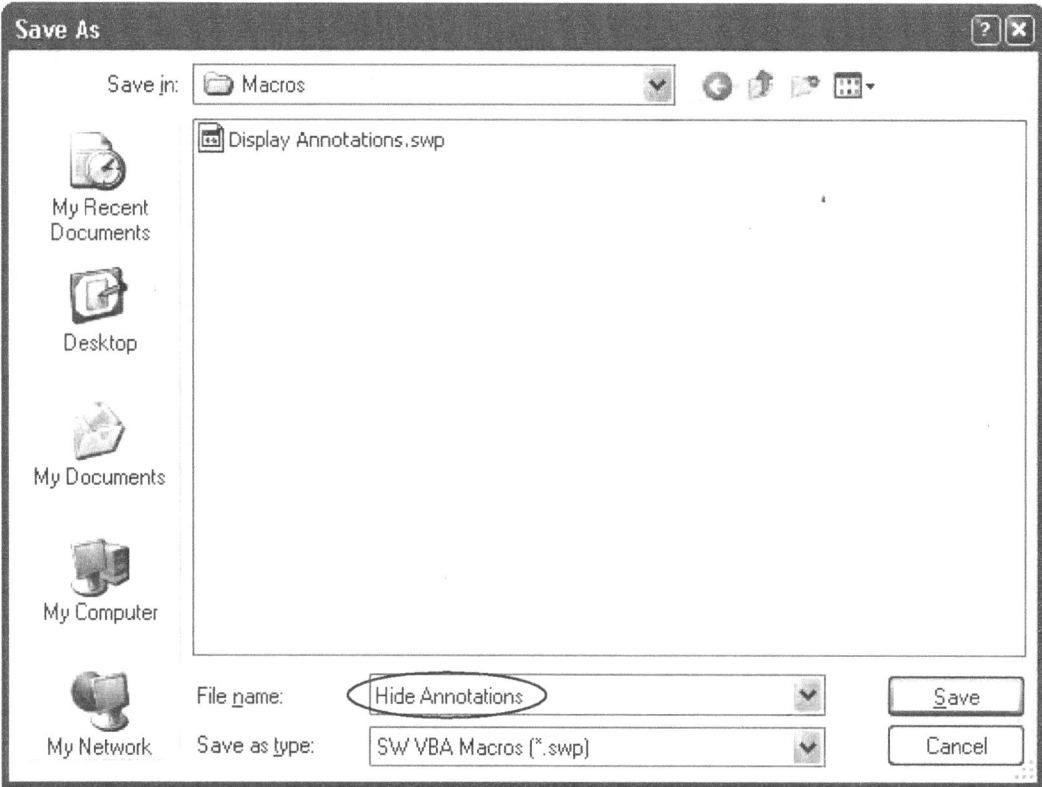

The dimensions are now hidden.

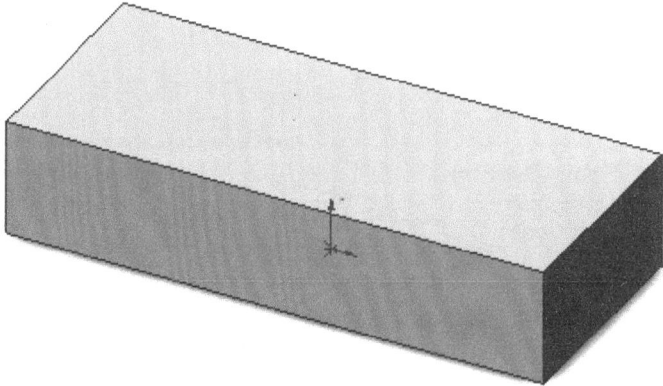

Run the Macros

▷ Click the **Run Macro** button on the "Macro" toolbar, or pull down the "Tools" menu and pick **Macro – Run**.

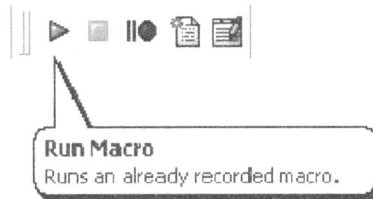

Run Macro
Runs an already recorded macro.

In the **Run Macro** dialog box, make sure that you are in the **Macros** folder, and pick **Display Annotations**.

Then, click **Open**. You could simply double click on **Display Annotations**.

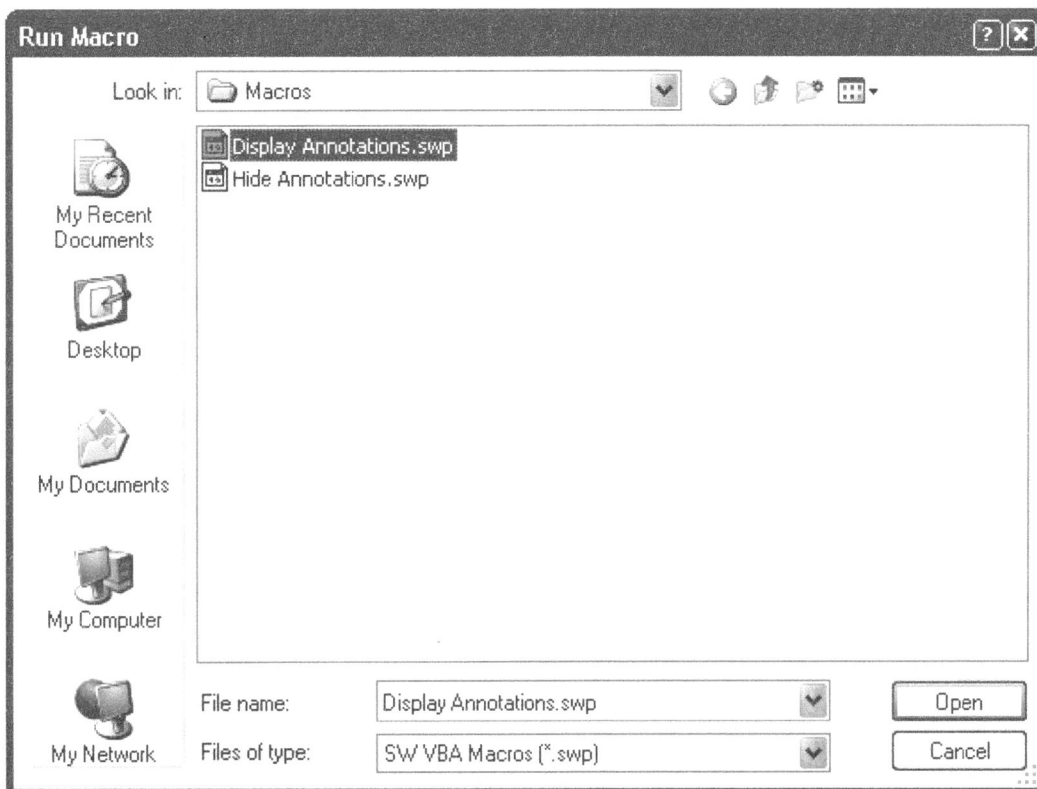

The macro is run and the dimensions appear on the screen. Just want we wanted to happen.

▷ Click the **Run Macro** button again on the "Macro" toolbar, or pull down the "Tools" menu and pick **Macro – Run**.

In the **Run Macro** dialog box, pick **Hide Annotations**, and click **Open**.

This macro is run and the dimensions are now hidden.

Create a Custom Macro Button

Pull down the "Tools" menu and pick **Customize**, or press **Ctrl+Shift+C**.

In the **Customize** dialog box, click on the **Commands** tab.

In the **Categories** list, click on **Macro**.

Under **Buttons**, click on and drag the **New Macro Button** button onto the "Macro" toolbar. Make sure that the cursor shows the plus sign next to it. Note that you can add a custom macro button to any toolbar.

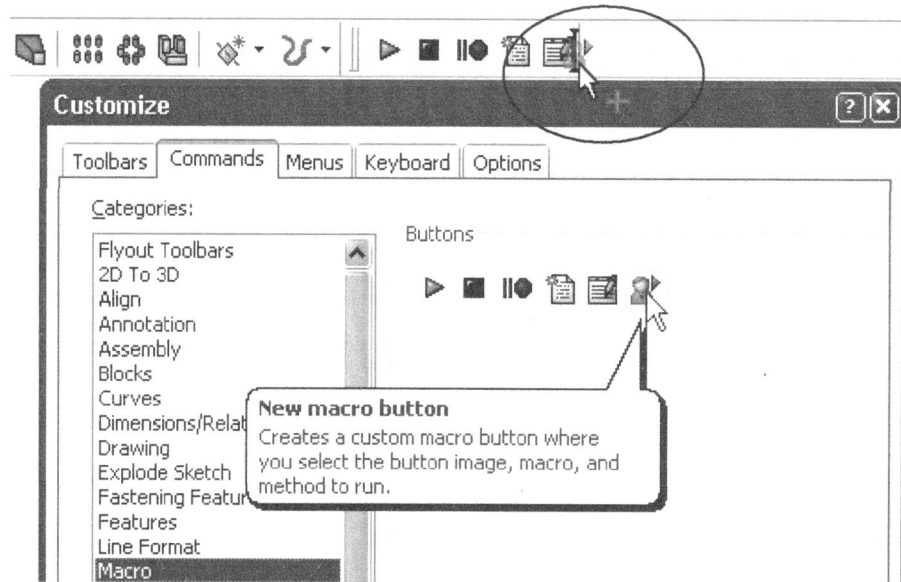

In the **Customize Macro Button** dialog box, under **Action**, click the `...` button.

In the **Macro path** dialog box, click on **Display Annotations.swp** in the **<SolidWorks installation directory>/Macros** folder, and then click **Open**.

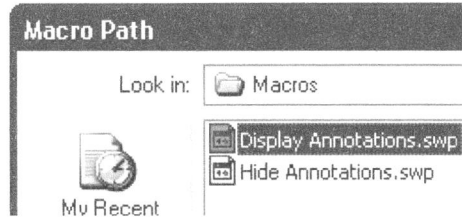

In the **Customize Macro Button** dialog box, under **Appearance**, click the **Choose Image** button.

In the **Icon path** dialog box, browse to the **<SolidWorks installation directory>/data/user macro icons** folder select **arrowup.bmp**. SolidWorks includes sample bitmaps, or you can create your own bitmap (.bmp) file. If you create your own bitmap, the bitmap must be 16 x 16 pixels, 256 colors, and the background color has to be white.

The best way to do this is to open one of the existing bitmap files in an image program like Microsoft Paint, and then use the **Save As** command to assign your name to the file. Then you can edit the image anyway you like and save it. Now, instead of selecting the **arrowup.bmp** above, you can select your own .bmp file.

Check **Preview** to see a preview of the bitmap image, and then click **Open**.

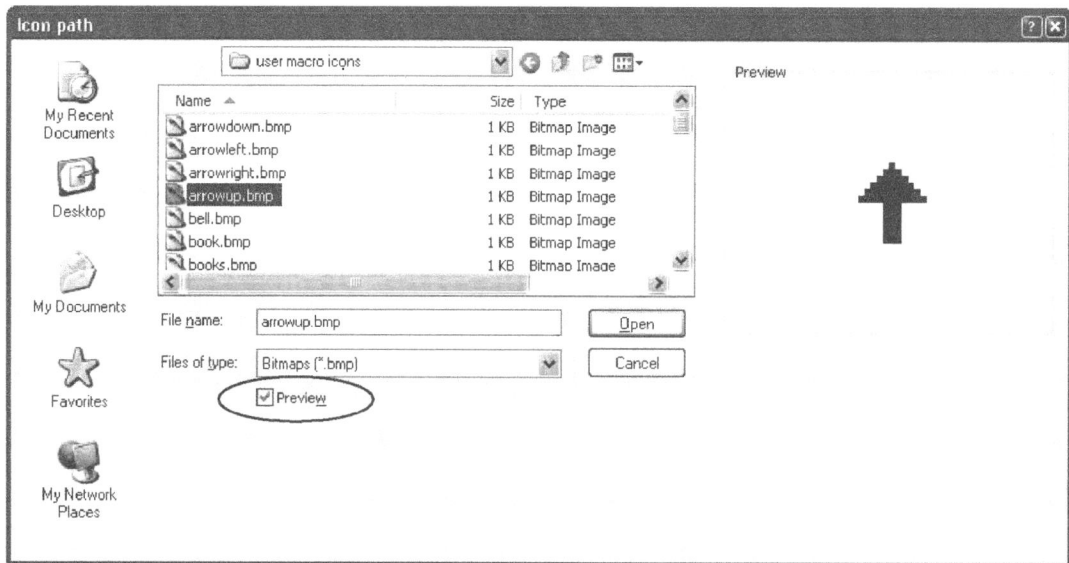

Click in the **Tooltip** box and type '**Display Annotations**'.

Click in the **Prompt** box and type '**Shows the dimensions of a model.**', which is shown on the status bar.

Click in the **Shortcut** box and press the **d** key to assign a shortcut key to the macro.

In the **Customize Macro Button** dialog box, click **OK**.

In the **Customize** dialog box, under **Buttons**, click on and drag another **New Macro Button** button onto the "Macro" toolbar.

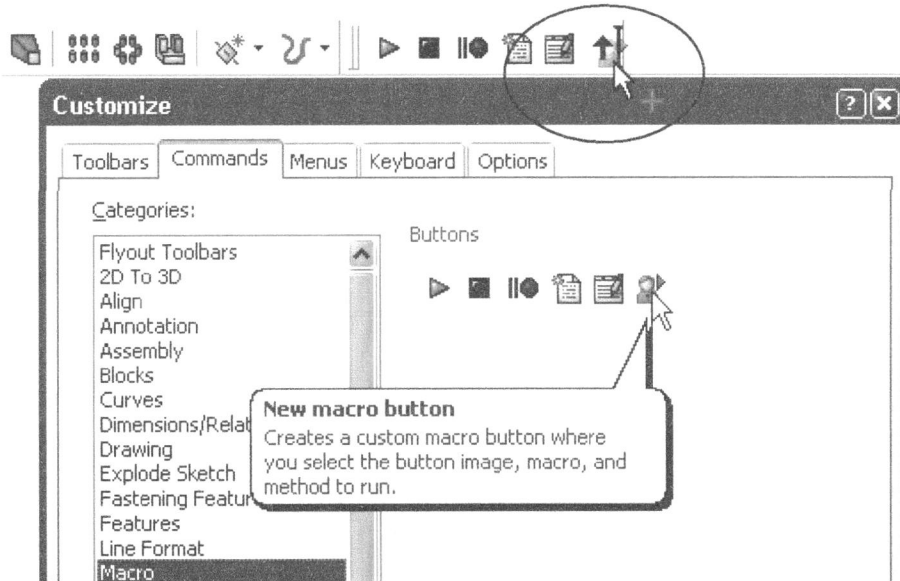

In the **Customize Macro Button** dialog box, under **Appearance**, click the **Choose Image** button.

In the **Icon path** dialog box, in the **<SolidWorks installation directory>/data/user macro icons** folder select **arrowdown.bmp**, and then click **Open**.

In the **Customize Macro Button** dialog box, click in the **Tooltip** box and type '**Hide Annotations**'.

Click in the **Prompt** box and type '**Hides the dimensions of a model.**'.

Under **Action**, click the [...] button.

In the **Macro Path** dialog box, browse to the **<SolidWorks installation directory>/Macros** folder and click on **Hide Annotations.swp**. Then click **Open**, or just double click on **Hide Annotations.swp**.

Click in the Shortcut box and press **Shift+d** to assign a shortcut key to the macro.

In the **Customize Macro Button** dialog box, click **OK**.

Click **OK** again to close the **Customize** dialog box.

Clicking on the new buttons in the "Macro" toolbar runs the selected macro.

Click on the **Display Annotations** button. The macro is run and the dimensions appear on the screen. Note the display of the tooltip when the cursor is placed over the button. Also notice that the prompt appears in the status bar in the lower left hand corner of the SolidWorks window.

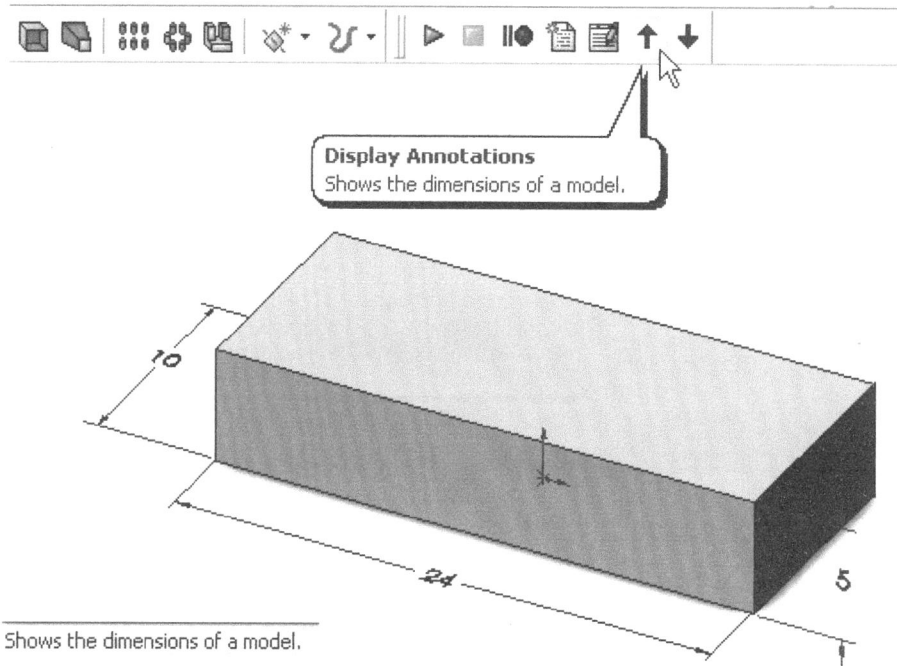

Shows the dimensions of a model.

Now, click on the **Hide Annotations** button. The macro is run and the dimensions are now hidden again.

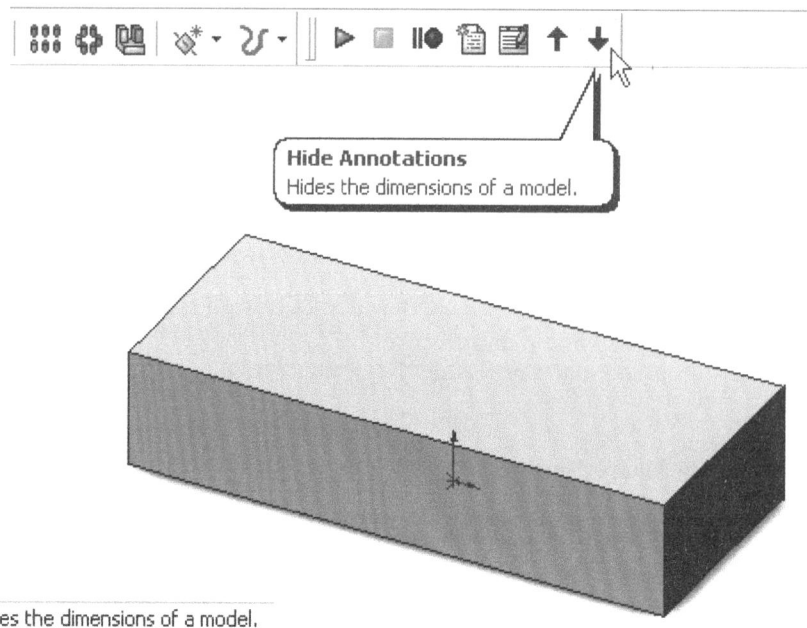

Hides the dimensions of a model.

Edit the Custom Macro Buttons

You can edit any of the settings used to create the custom macro button. You can also change where the button is and which toolbar it is located on.

First, click the **Explode Sketch** button in the control area of the CommandManager to display the custom toolbar in the CommandManager.

Pull down the "Tools" menu and pick **Customize**, or press **Ctrl+Shift+C**.

With the **Customize** dialog box showing, simply right click on the **Display Annotations** button in the "Macros" toolbar.

The **Customize Macro Button** dialog box appears. You can make any changes to the macro button settings.

Click in the **Prompt** box and change the text to '**Displays the dimensions of a model.**', and then click **OK**.

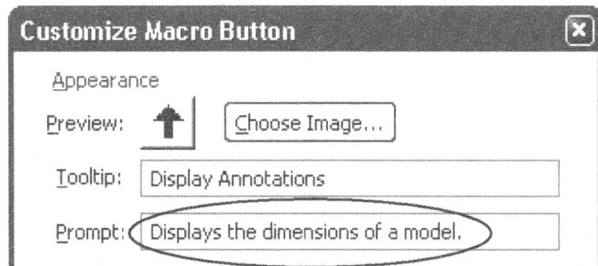

Now, drag the **Display Annotations** button onto the CommandManager between the **Open** button and the **Sketch** button. Be careful! If you drag the button off a toolbar, the button will be deleted and must be recreated from scratch. Also note that you can only customize a macro button from a toolbar, not from the CommandManager.

When the **Customize Macro Button** dialog box appears, click **OK**.

Drag the **Hide Annotations** button onto the CommandManager between the **Display Annotations** button and the **Sketch** button.

When the **Customize Macro Button** dialog box appears, click **OK**.

Click **OK** again to close the **Customize** dialog box.

The CommandManager should now include the two new buttons.

Don't forget that **D** and **Shift+D** are still the assigned keyboard shortcut keys. You can use those instead of clicking on the custom buttons.

Assign a Macro to a Shortcut Key

You don't have to create a custom button if you only want a keyboard shortcut key assigned to your macro. After you create a **Macros** folder and record a macro, you can simply assign the macro to a shortcut key.

On the "Macro" toolbar, click the **Record\Pause Macro** button, or pull down the "Tools" menu and pick **Macro – Record**.

Press **Ctrl+B** to rebuild the model.

Next, press **Ctrl+7** to rotate the model to the isometric view orientation.

Then, press the **F** key to zoom the model to fit the window.

Click the **Stop Macro** button on the "Macro" toolbar, or pull down the "Tools" menu and pick **Macro – Stop**.

In the **Save As** dialog box, browse to the **<SolidWorks installation directory>/Macros** folder.

Click in the **File name** box and type '**Isometric Rebuild**', and then click **Save**.

Pull down the "Tools" menu and pick **Customize**, or press **Ctrl+Shift+C**.

In the **Customize** dialog box, click on the **Keyboard** tab.

Pull down the **Category** list and pick **Macros**. **Macros** is listed because you created a **Macros** folder in your SolidWorks installation directory and you saved your macros there.

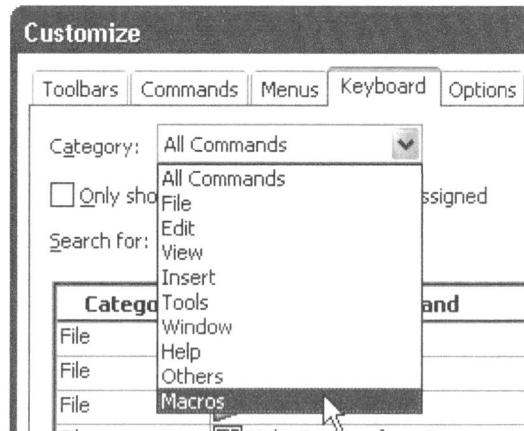

Click in the **Shortcut(s)** box next to the **Isometric Rebuild** command.

Press **Ctrl+Shift+B** for the shortcut key, and then click **OK**.

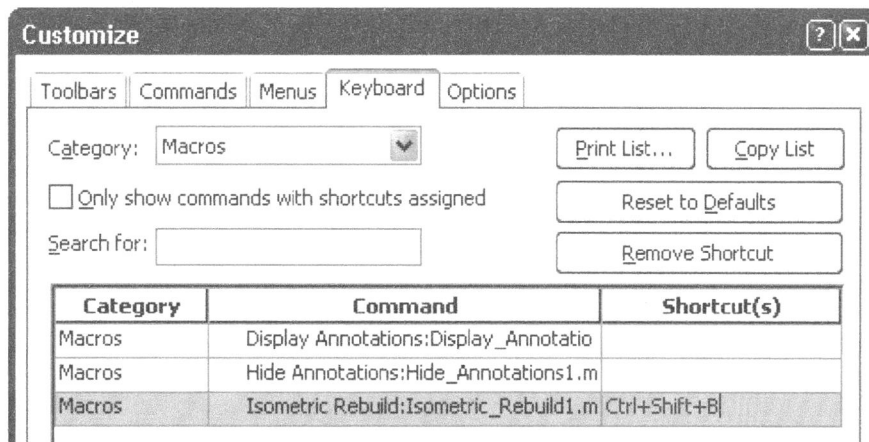

Press **D** to show the dimensions, and then, double click the **24** dimension and change it to **20** and press **Enter**.

Press **Shift+D** to hide the dimensions. Note that the model has not been rebuilt to the **20** dimension yet.

Press **Ctrl+Shift+B** to rebuild the model and change to the Isometric view orientation. Now, you can use this macro very easily whenever you want.

Finally, hide the "Macro" toolbar by right clicking on the "Macro" toolbar and picking **Macro** from the list to deselect it and hide the toolbar.

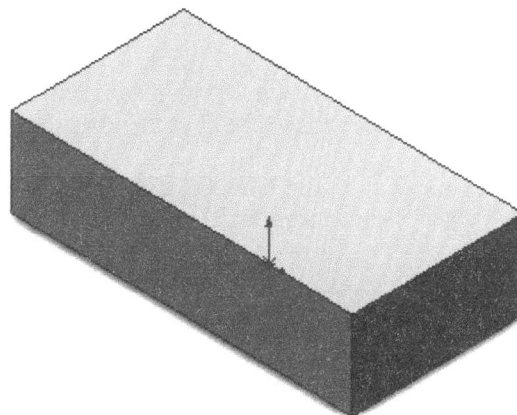

Edit the Macro

There is a lot more you can do with a macro. But to get there you must be familiar with Visual Basic. This exercise adds a message box to the **Isometric Rebuild** macro you created above. It uses an **IF** statement to only change the view if you pick **Yes.** Otherwise, skip the view change if you pick **No**.

Click the **Edit Macro** button on the "Macro" toolbar, or pull down the "Tools" menu and pick **Macro – Edit**.

The **Edit Macro** dialog box appears. You may need to browse to the correct folder where your macros are saved. Pick the **Isometric Rebuild.swp** file, and then click **Open**.

SolidWorks opens the Microsoft Visual Basic program and displays the lines of code for this macro. This is what a recorded macro looks like. Now you see why you need to know what you are doing before you attempt this on your own.

```
Dim swApp As Object
Dim Part As Object
Dim SelMgr As Object
Dim boolstatus As Boolean
Dim longstatus As Long, longwarnings As Long
Dim Feature As Object
Sub main()

Set swApp = Application.SldWorks

Set Part = swApp.ActiveDoc
Set SelMgr = Part.SelectionManager
boolstatus = Part.EditRebuild3
Part.ShowNamedView2 "*Isometric", 7
Part.ViewZoomtofit2
End Sub
```

Click with your cursor below the line which states **Sub main()** and type '**Dim answer**'. This creates a variable for you to save the answer to the question you will ask.

Next, click with the cursor in front of the line that says **Part.ShowNamedView2 "*Isometric", 7** and press the **Enter** key to create a new line. Then, type these two new lines:

> **answer = MsgBox("Change to Isometric View?", 4, "View")**
> **If answer = 6 Then**

The first line will display a message box titled **View** and ask the question **Change to Isometric view?** The number 4 indicates that you want **Yes** and **No** buttons. The second line tells the macro that if the **Yes** button was picked, then do the commands which follow.

One more thing that you need to do is mark the end of the commands to be performed as part of the **IF** statement. Click with the cursor in front of the line that says **End Sub** and press the **Enter** key to create a new line. Type a new line which says '**End If**'.

Your code should now look like what is shown here.

Pull down the "File" menu and pick **Close and Return to SolidWorks**.

SolidWorks will ask if you want to save the changes to your macro. Click the **Yes** button.

Now, try running the macro by pressing **Ctrl+Shift+B**.

After it rebuilds the model, it displays your message asking if you want to **Change to Isometric View?** If you click the **No** button, the view does not change. If you click the **Yes** button, the view changes to the Isometric view.

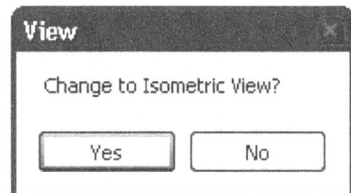

```
Dim swApp As Object
Dim Part As Object
Dim SelMgr As Object
Dim boolstatus As Boolean
Dim longstatus As Long, longwarnings As Long
Dim Feature As Object
Sub main()

Dim answer

Set swApp = Application.SldWorks

Set Part = swApp.ActiveDoc
Set SelMgr = Part.SelectionManager
boolstatus = Part.EditRebuild3
answer = MsgBox("Change to Isometric View?", 4, "View")
If answer = 6 Then
        Part.ShowNamedView2 "*Isometric", 7
        Part.ViewZoomtofit2
End If
End Sub
```

View

Change to Isometric View?

[Yes]   [No]

Once again, there is a lot you can do with programming macros through Visual Basic. But you will want to learn much more before you attempt it on your own.

Chapter 7

Design Library

The Design Library is a great concept, centrally locating reusable, drag-and-drop features, annotations, notes, blocks, parts and assemblies. Located in the Task Pane, the Design Library not only speeds up design time, it helps to maintain company standards. Although the Design Library is easy to find and use, the default items that you may want to use regularly are scattered throughout, or you will have to create them yourself. They are your standard parts after all. In this chapter, you will see how easy it is to modify the Design Library by creating your own folders and placing your favorite tools in them.

Commonly used features, such as pockets or slots, can be created and saved as library features. This provides great time savings and promotes standardization by making the geometry consistent among your CAD files. Think about selecting the desired feature from a list rather than drawing it new each time you need it.

When working in a multi-user environment, you should put the Design Library on a shared network server accessible by all. If moving a library, be aware of location references. If you are putting a part or assembly into the Design Library, do not save it in the normal area where you save your documents. Save it only in the library folder through the Save dialog. This will eliminate duplicate files.

Access the Design Library

With the Task Pane shown on the right hand side of your screen, click the **Design Library** tab in the Task Pane. Remember, to expand or collapse the Task Pane, click an arrow or anywhere along the bar between the arrows. If you have never been here before, open several of the folders and look around the default Design Library. SolidWorks installs a few sample library features, annotations, parts, assemblies, and so forth. The Toolbox, only available in SolidWorks Office Professional & Premium, contains standard "Machinery's Handbook" parts and components. 3D ContentCentral® contains a free online vendor supplied component directory as well as an on-line User Library to share models and assemblies with other SolidWorks users (requires an internet connection).

The **Design Library** tab in the Task Pane provides a central location for reusable elements such as features, parts, and even forming tools.

Design Library
Click to display this task pane tab.

In the **Design Library** tab, click the **Add File Location** button. This is the easiest way to create a new Design Library location.

In the **Choose Folder** dialog box, browse to the installation folder of SolidWorks, usually in the Program Files folder on your local hard drive. Then, double click on the **data** folder.

While in the **data** folder, in the **Choose Folder** dialog box, click the **Create New Folder** button.

A new folder is created in the **data** folder. Enter '**My Library**' for the folder name. You may assign any valid Windows folder name you wish, such as a reference to your company, or the type of parts or features you plan to save in this folder.

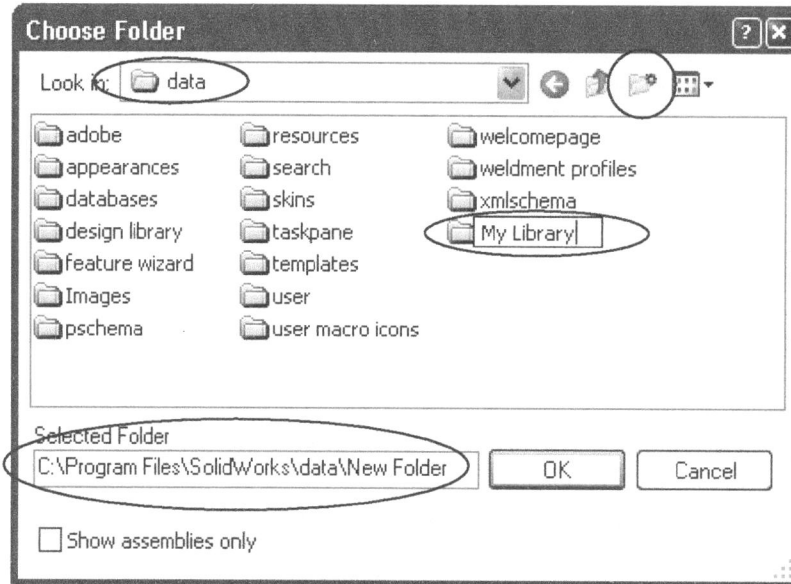

Double click on the **My Library** folder icon and then click the **OK** button.

If the **3D ContentCentral Terms of Use** dialog box appears, click **Accept**.

The **My Library** Design Library has now been created.

In the **Design Library** tab, right click on **My Library** and pick **New Folder** from the menu. You may also click on **My Library** to highlight it and then click on the **Create New Folder** button.

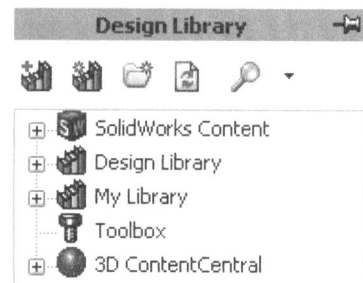

A new folder is created under **My Library**. Enter '**My Features**' for the folder name. I found this to be the best and most efficient way to create folders to help organize the items in the Design Library.

Copy an Existing Library Feature

A library feature is a frequently used feature, or combination of features, that you create once and then save in a library for future use. To help you get familiar with library features, SolidWorks has provided sample features that you can use or customize.

In the **Design Library** tab, click on the plus sign next to the **Design Library** folder to expand the folder. Then, click on the plus sign next to the **features** folder to expand the folder. Click on the plus sign next to **metric**.

Click on the **hole patterns** folder in the upper pane to view the contents of the folder in the lower pane.

Scroll the upper pane so that you can see your **My Features** folder.

In the lower pane, left click and drag the **linear hole pattern**. Hold down the **Ctrl** key to copy the feature (a plus sign should appear attached to the cursor as shown). Move the cursor over the **My Features** folder and release the mouse button. If you accidentally move the feature without copying it, click on the folder you dragged it into. Then, re-drag the file while holding the **Ctrl** key to copy it back to the original folder. You may need to scroll the upper pane so that you can see the original folder.

In the upper pane in the **features** folder, under the **metric** folder, click on the **slots** folder to view the contents of this folder in the lower pane.

Scroll the upper pane so that you can see your **My Features** folder. Left click and drag **straight slot** while holding the **Ctrl** key and drop it onto the **My Features** folder.

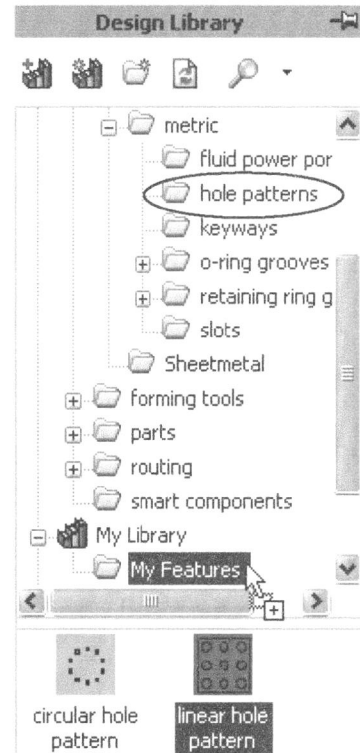

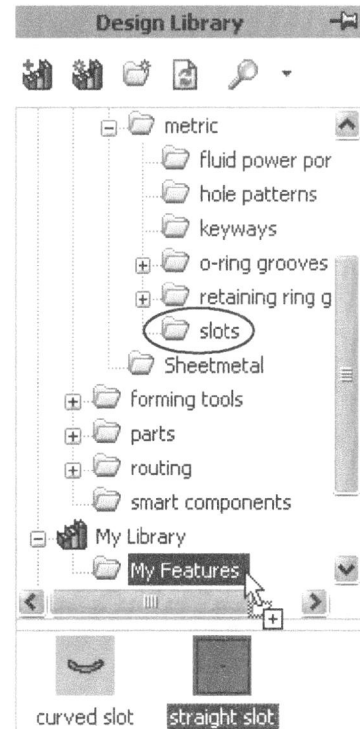

In the upper pane, click on the **My Features** folder to view the contents of this folder in the lower pane.

Right click on **straight slot** and pick **Rename** from the menu. Type '**metric slot**' for the file name. The file names follow the standard for Windows. The complete path and file name and extension can contain up to 255 characters. File names cannot include a forward slash (/), backslash (\), greater than sign (>), less than sign (<), asterisk (*), question mark (?), quotation mark ("), pipe symbol (|), or colon (:).

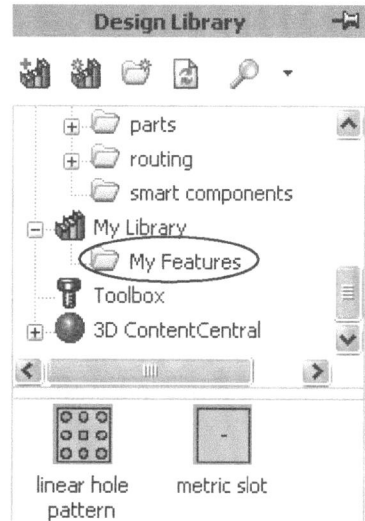

Create a Custom Feature

To create a new library feature, you first create a base feature to which you add the features that you want included in the library feature.

Create a new **Part** file by clicking the **New** button in the "Standard" toolbar, or pull down the "File" menu and pick **New**.

Click the **Explode Sketch** button in the control area of the CommandManager. Then, click the **Extruded Boss/Base** button from the toolbar, or pull down the "Insert" menu and pick **Boss/Base – Extrude**.

Select the **Top** plane when prompted to select a plane.

Create a rectangle with the lower left corner at the origin using the **Rectangle** button in the CommandManager, or pull down the "Tools" menu and pick **Sketch Entities – Rectangle**.

Click the **Smart Dimension** button in the CommandManager, or pull down the "Tools" menu and pick **Dimensions – Smart**.

Add a '**35**' vertical dimension to the left vertical line and a '**35**' horizontal dimension to the bottom horizontal line.

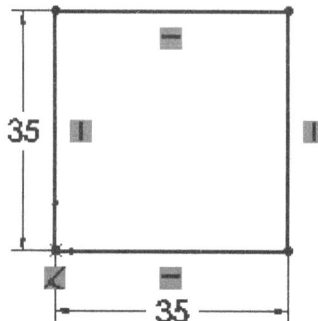

Exit the sketch by clicking the **Exit Sketch** button in the CommandManager or in the upper right corner of the graphics area.

In the **Extrude** PropertyManager, under **Direction 1**, set the **Depth** to '**5**'.

Click the green check mark button at the top of the **Extrude** PropertyManager to accept the settings and create the part.

Click the **Extruded Cut** button from the CommandManager or pull down the "Insert" menu and pick **Cut – Extrude**.

Select the top of the part as the plane onto which you will create the sketch.

In the bottom left corner of the graphics area, change the View orientation by clicking the pull down arrow and picking **Top**, or press **Ctrl+5**.

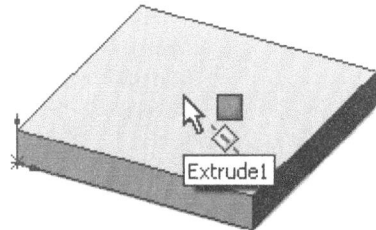

Create a rectangle in the middle of the part using the **Rectangle** button in the CommandManager, or pull down the "Tools" menu and pick **Sketch Entities – Rectangle**.

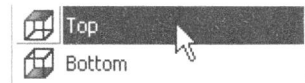

Right click in the graphics area and pick **Smart Dimension** from the menu, and add a '**25**' dimension to both sides of the rectangle. You do not want to dimension this new rectangle relative to the base rectangle so that there will not be any relations between them. This will give you more flexibility when you place it into the part. It doesn't have to be centered here, just make certain the rectangle is completely on the base.

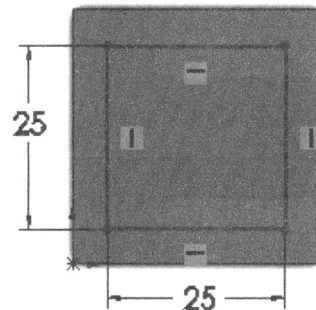

Exit the sketch by clicking the **Exit Sketch** button in the CommandManager or in the upper right corner of the graphics area.

In the **Cut-Extrude** PropertyManager, under **Direction 1**, set the **End Condition** to **Through All**, and then click the green check mark button at the top of the **Cut-Extrude** PropertyManager.

Pull down the "Tools" menu and pick **Options**.

On the **Document Properties** tab, pick **Colors**.

Scroll down the **Model\Feature colors** box and pick **Library Feature**.

Click on the **Edit** button. In the **Color** dialog box, pick the yellow color (second column, second row) and click **OK**. In the **Document Properties** dialog box, click the **OK** button.

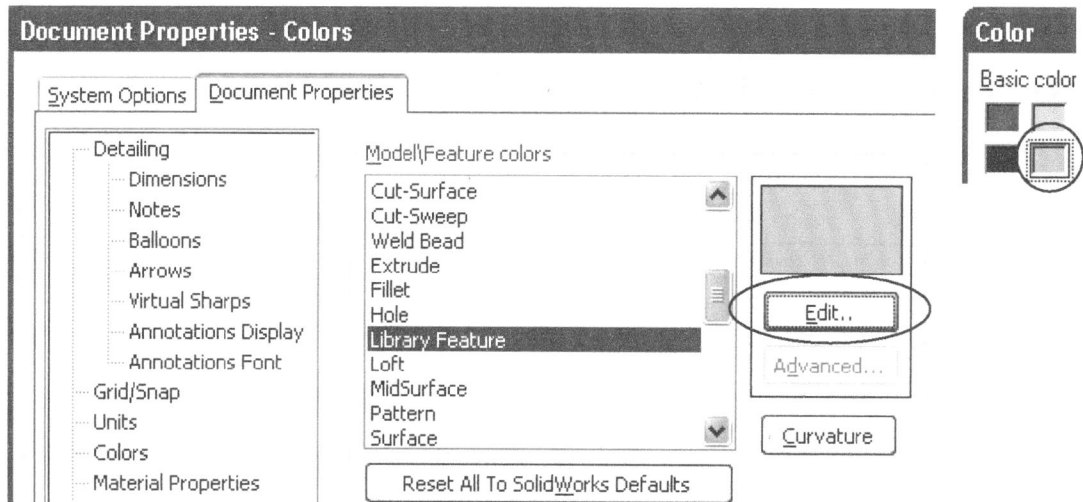

Each time you add the library feature to a part, the applied color is displayed.

Click the **Design Library** button in the Task Pane.

Click on the **Pin** button in the title bar of the Task Pane to keep the Task Pane open while you are working. If the Task Pane is unpinned, it collapses when you drag an item into the graphics area, or when you open a new SolidWorks document. You can click on the **Pin** button at any time to unpin the Task Pane.

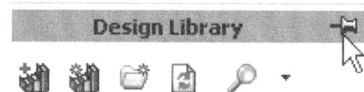

When you save a library feature or a forming tool, the thumbnail graphic reflects the view when the document is saved. Before saving, zoom in and orient the part or assembly so that the thumbnail preview that appears in the lower pane of the Design Library will look the way that you want it to look. The Top view of the rectangle should work just fine.

Make sure that the **My Features** folder is displayed in the Design Library. Then, in the FeatureManager design tree, click and drag **Cut-Extrude1** to the lower pane of the **Design Library** tab. The cursor will change as shown, indicating that you are copying the feature. Release the mouse button.

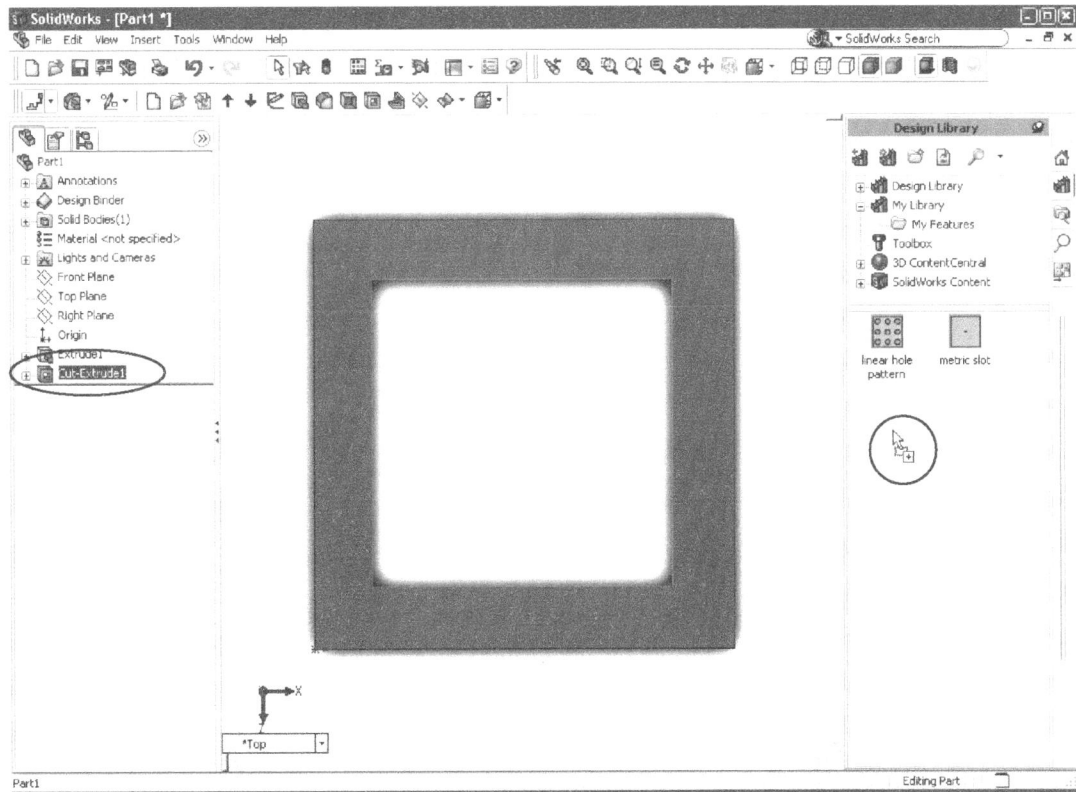

In the **Add to Library** PropertyManager, under **Save To**, type '**25 x 25 Rect**' for the **File name**.

Under **Design Library folder**, make sure that **My Features** is highlighted.

Under **Options**, make sure **File type** is set to **Lib Feat Part (*.sldlfp)**. Library features have the .sldlfp extension.

Click the green check mark button at the top of the **Add to Library** PropertyManager to accept the settings and create the library feature.

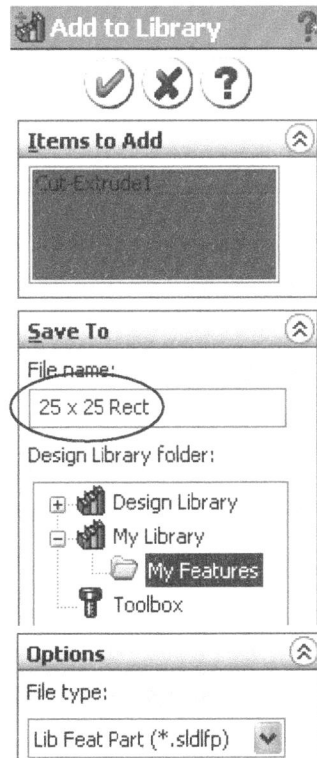

Click on the lower **X** button in the upper right corner of the SolidWorks window to close the file, or pull down the "File" menu and pick **Close**.

In the **SolidWorks** dialog box, click **No** when asked to save changes to the part.

In the Design Library, click on the **My Features** folder. Click the **Refresh** button to make certain everything is up to date. You should now have three features in this folder.

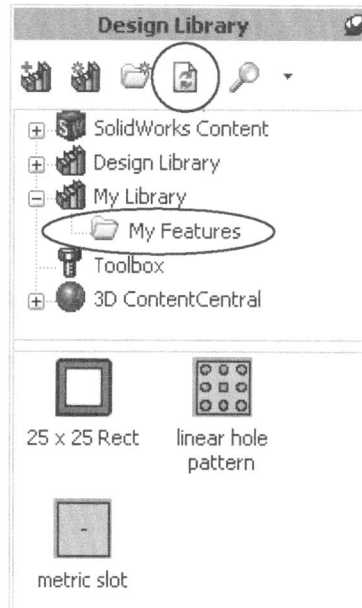

Edit a Custom Feature

You can easily edit an existing library feature.

In the Design Library, in the **My Features** folder, right click on **25 x 25 Rect** and pick **Open**.

Click on the **FeatureManager design tree** tab, and then right click on **Cut-Extrude1** and pick **Edit Sketch**.

Add fillets with a radius of '**5**' on all four corners of the rectangle using the **Sketch Fillet** button in the CommandManager, or pull down the "Tools" menu and pick **Sketch Tools – Fillet**.

Click the green check mark button at the top of the **Sketch Fillet** PropertyManager.

Exit the sketch by clicking the **Exit Sketch** button in the CommandManager or in the upper right corner of the graphics area.

In the FeatureManager design tree, right click on **Cut-Extrude1** and pick **Edit Feature**.

In the **Cut-Extrude1** PropertyManager, under **Direction 1**, change **Through All** to **Blind** and enter a **Depth** of '**4**'.

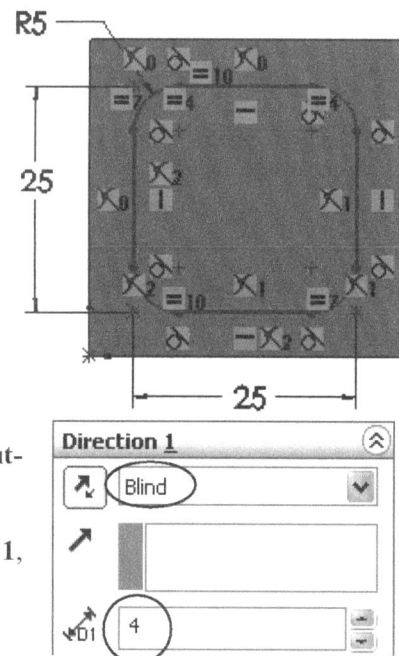

Click the green check mark button at the top of the **Cut-Extrude1** PropertyManager.

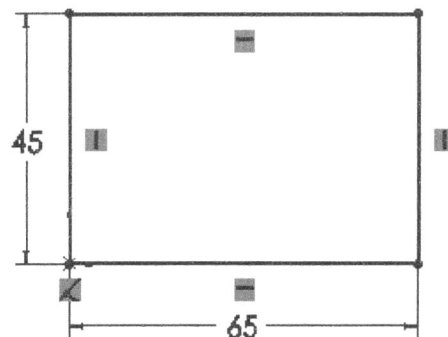

Press **Ctrl+S** or pull down the "File" menu and pick **Save**. The bottom of the pocket should be yellow.

Click on the lower **X** button in the upper right corner of the SolidWorks window to close the file, or pull down the "File" menu and pick **Close**.

In the Design Library, under the **My Features** folder, right click on **25 x 25 Rect** and pick **Rename**.

Change **25 x 25 Rect** to '**25 x 25 Pocket**' as shown and press the **Enter** key.

In the Design Library, click on the **Refresh** button to update the thumbnail preview of the 25 x 25 Pocket.

Add a Library Feature to a Part

You can add a library feature to a part by simply dragging the library feature from the Design Library to the part. The library feature will be the color that you set in the Document Properties.

To do this, create a new **Part** file by clicking the **New** button in the "Standard" toolbar, or pull down the "File" menu and pick **New**.

Click the **Explode Sketch** button in the control area of the CommandManager. Then, click the **Extruded Boss/Base** button from the toolbar, or pull down the "Insert" menu and pick **Boss/Base – Extrude**.

Select the **Front** plane when prompted to select a plane.

Create a '**45**' by '**65**' rectangle with the lower left corner at the origin using the **Rectangle** button in the CommandManager, or pull down the "Tools" menu and pick **Sketch Entities – Rectangle**.

Exit the sketch by clicking the **Exit Sketch** button in the CommandManager or in the upper right corner of the graphics area.

In the **Extrude** PropertyManager, under **Direction 1**, set the **Depth** to '**35**' and click the green check mark button at the top of the **Extrude** PropertyManager.

In the Design Library, under the **My Features** folder, click and drag **25 x 25 Pocket** onto the front of the block you just created, as shown.

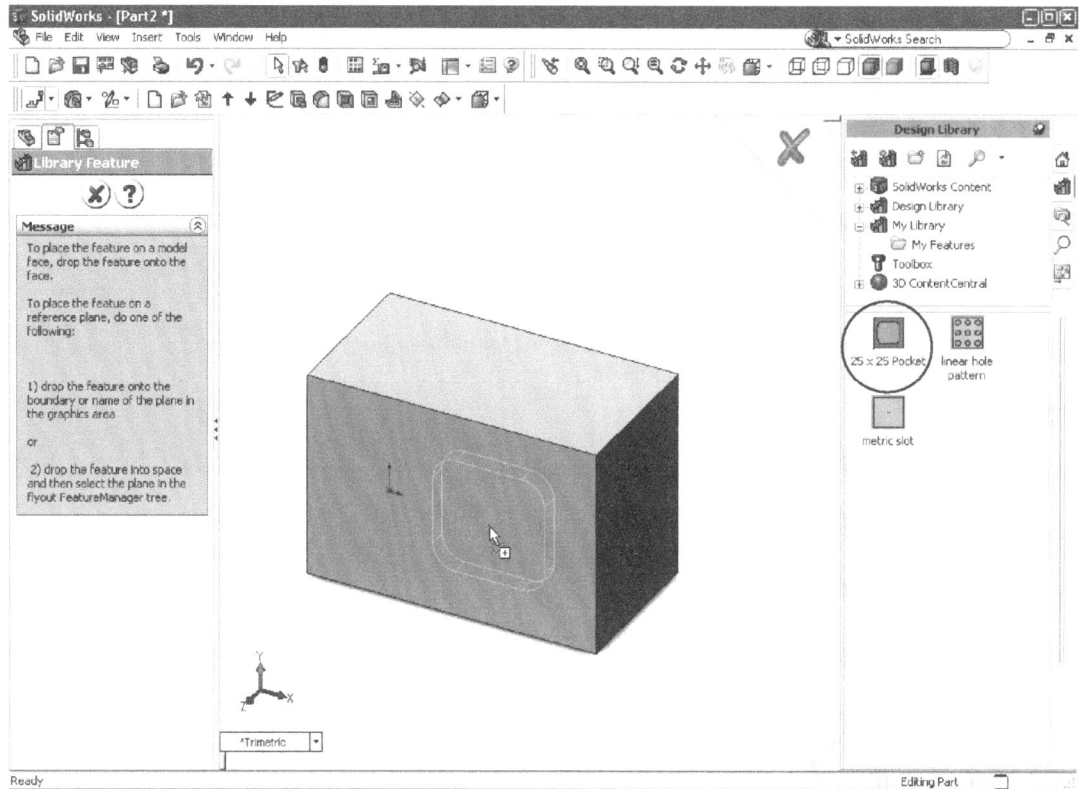

In the **25 x 25 Pocket** PropertyManager, check **Link to library part** so the library feature you inserted will automatically update if you ever edit the original library feature. You can also edit the sketch to locate the pocket.

Do not override the dimension values, and click the green check mark button at the top of the **25 x 25 Pocket** PropertyManager.

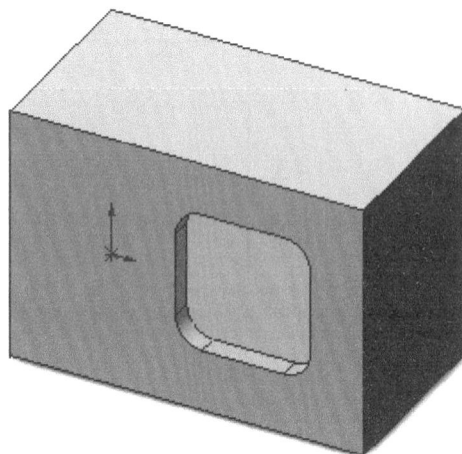

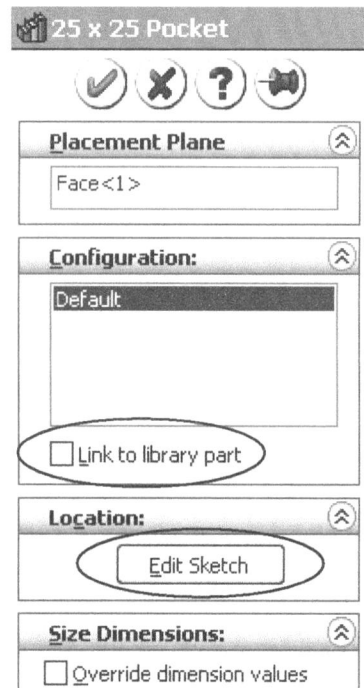

Modify the Design Library

You can customize the Design Library to fit your needs, using folders and subfolders to organize your parts and features.

Pull down the "Tools" menu and pick **Options**.

On the **System Options** tab, pick **File Locations**.

Under **Show folders for:**, pull down the list and pick **Design Library**. The default Design Library folder is located in the **<SolidWorks Install Directory>/data/Design Library**.

Under **Folders:**, pick **<SolidWorks Install Directory>\data\My Library**, and then click the **Move Up** button. This moves **My Library** above **Design Library** in the Task Pane.

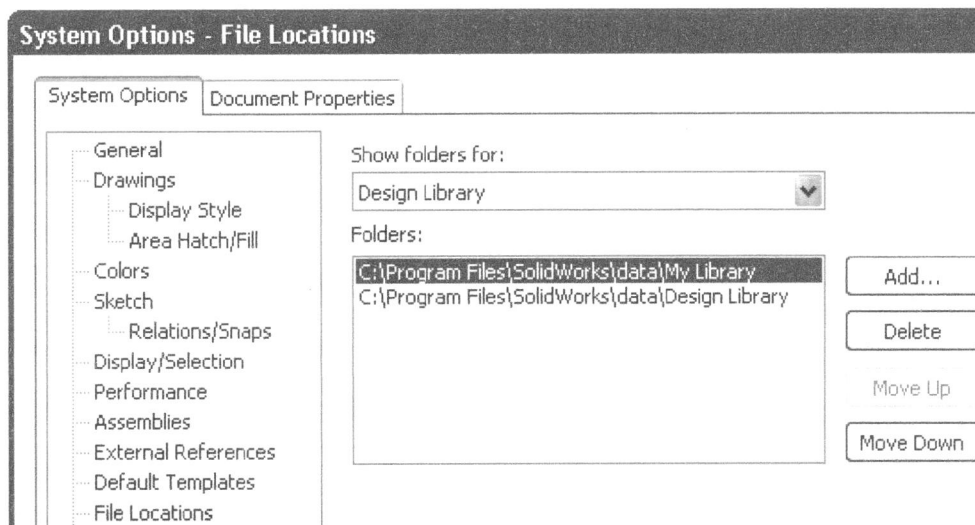

Click **OK**. You have this ability to control the order of your Design Libraries whenever you want.

The best way to remove or delete a folder that you no longer want is just to right click on the folder in the Design Library and pick **Delete** from the menu.

In the Design Library, right click on **My Features** and pick **Rename**. Type 'Features' and press the **Enter** key. Changing the folder names can be done at any time.

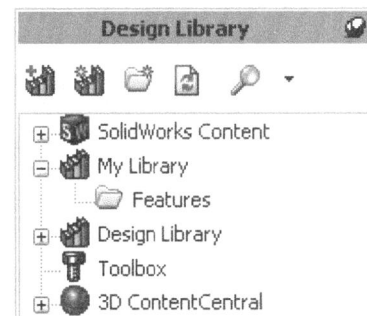

Click the **Pin** button in the title bar of the Task Pane to unpin the Task Pane.

Pull down the "File" menu and pick **Close**. It is up to you if you want to save the part. We will not be using it anymore.

Create Your Own Design Library

Now take what you've learned and think about what things you will want to place in your Design Library. Then think about the folder structure and how it will be organized to benefit you the most. The Design Library is a great tool to organize your commonly used features, notes, blocks, parts, and assemblies. Apply this chapter to create a custom Design Library that will help you be more productive.

Maybe you have a feature like the previous pocket that appears on many of your parts. Or maybe you need quick access to standard components or parts when building an assembly. Perhaps you have a set of standard notes that you frequently place on your drawings. Take all of these factors into consideration when customizing your personal Design Library. Remember, too, that you can change and modify how things are organized whenever you want. SolidWorks has a default directory structure that may make sense to you, or maybe you want to create your own. The program allows so much flexibility with this great tool that I can't even begin to tell you what will work best for you. I added some new folders, in this chapter, just to give you an idea of what can be done. It is now up to you to implement these ideas. Just try some things out and modify it after you gain some experience. Get together with the other users in your organization. There should really be only one Design Library shared by all of your users.

There is one big concern to be aware of though. It is very easy to create duplicate files without realizing it. For example, if you add an assembly or a part to the Design Library, the original file remains right where it was and a copy of the file is placed in the Design Library folder that you indicate. Watch for this and delete the original file to avoid the duplicates. Failure to do so will cause a lot of trouble. Edits will be made to one and not the other. At some point someone will use the wrong one.

SolidWorks allows you to drag and drop from the Design Library to the Graphics Area. You can also drag and drop from the Graphics Area, the FeatureManager design tree, or from Microsoft Windows Explorer to the Design Library. Dragging and dropping from folder to folder is also allowed. But this too will create duplicate files.

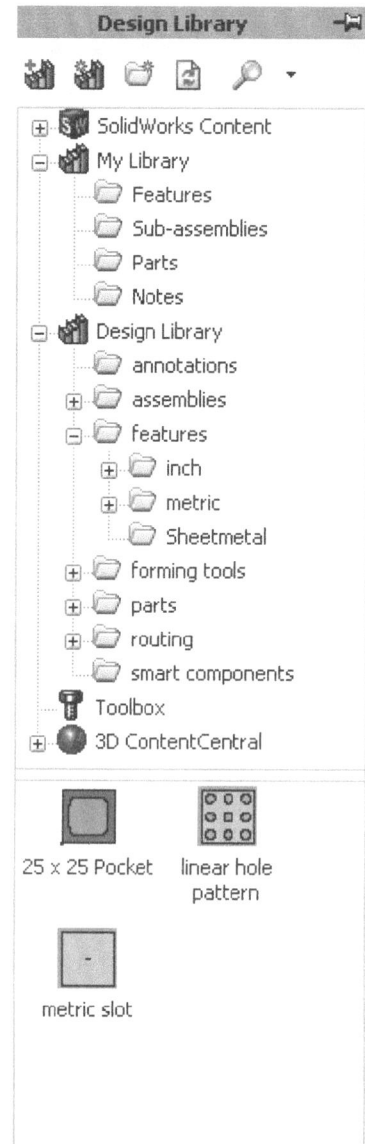

Chapter 8

SolidWorks allows you to rotate and zoom the model or drawing to a preset view. There are many standard views to choose from. Sometimes, though, you want to create a custom view and return to it later as you model your part. SolidWorks allows you to add your own named view to the list of standard views. You can also create a custom view that can be saved to the standard view orientation list.

Multiple views of the same document can be visible at the same time. The graphics area can be divided horizontally and vertically using the Split option. Each window is an independent view of your model, allowing you to see a part in two or four views simultaneously. SolidWorks also allows you to view more than one document window of an open document at a time. The new windows are simply additional views of the same part. Changes to the model in one view appear in all of the views.

Create a Base Feature

To start this chapter, create a new **Part** file by clicking the **New** button in the "Standard" toolbar, or pull down the "File" menu and pick **New**.

Click the **Explode Sketch** button in the control area of the CommandManager. Then, click the **Extruded Boss/Base** button from the toolbar, or pull down the "Insert" menu and pick **Boss/Base – Extrude**.

Select the **Front** plane when prompted to select a plane on which to sketch the feature cross-section.

Create a rectangle with the origin at the lower left corner using the **Rectangle** button in the CommandManager, or pull down the "Tools" menu and pick **Sketch Entities – Rectangle**.

Create an angled line as shown using the **Line** button in the CommandManager, or pull down the "Tools" menu and pick **Sketch Entities – Line**.

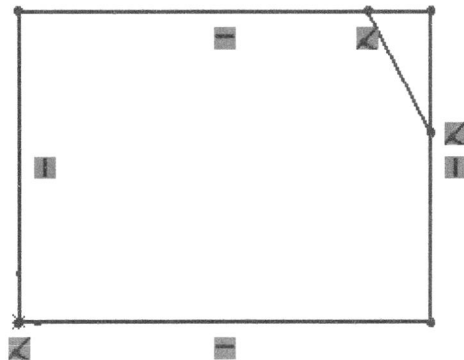

Trim the upper right hand corner using the **Trim Entities** button, or pull down the "Tools" menu and pick **Sketch Tools – Trim**.

Add dimensions to the sketch as shown here using the **Smart Dimension** button in the CommandManager, or pull down the "Tools" menu and pick **Dimensions – Smart.**

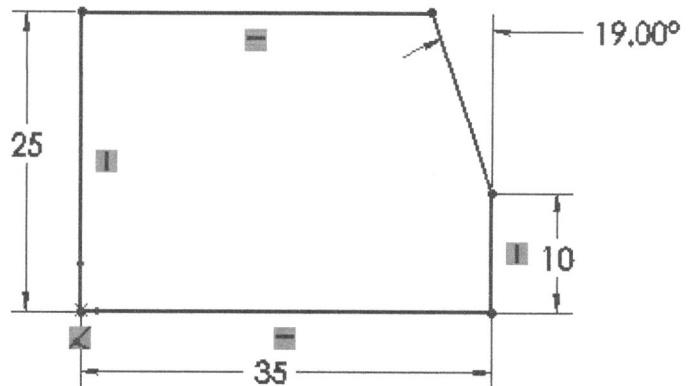

Exit the sketch by clicking the **Exit Sketch** button in the CommandManager or in the upper right corner of the graphics area.

In the **Extrude** PropertyManager, under **Direction 1**, set the **Depth** to '20'.

Click the green check mark button at the top of the **Extrude** PropertyManager to accept the settings and create the part.

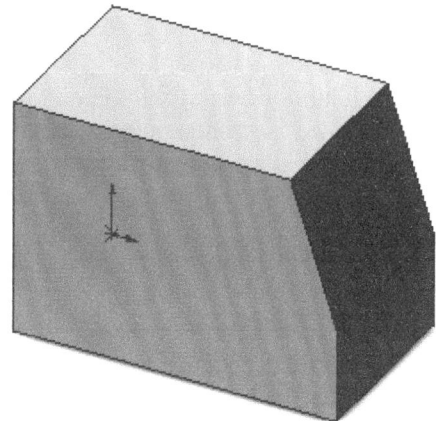

Orientation Dialog Box

The **Orientation** dialog box contains a list of the views on the "Standard Views" toolbar and any custom user views.

Pull down the "View" menu and pick **Modify – Orientation**, or just press the **Space Bar**.

To keep the **Orientation** dialog box open, click the **Pin** button in the dialog box. You may drag the dialog box wherever you like.

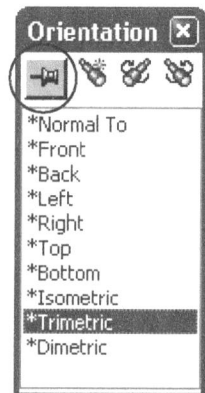

Change the Orientation of the Standard Views

The above part was created so that you can see the difference between the views. In the **Orientation** dialog box, double click on **Front** to change the orientation to the Front view.

Once in awhile you get a part that just isn't oriented the way you think it should be. What you think is the Top view is really the Front view. You can change this without rotating the model. For example, single click on **Top** in the **Orientation** dialog box to pick **Top**, the name of a standard view you want to assign to the current Front view orientation of the model.

Click the **Update Standard Views** button to update all of the standard views so they are relative to the selected view.

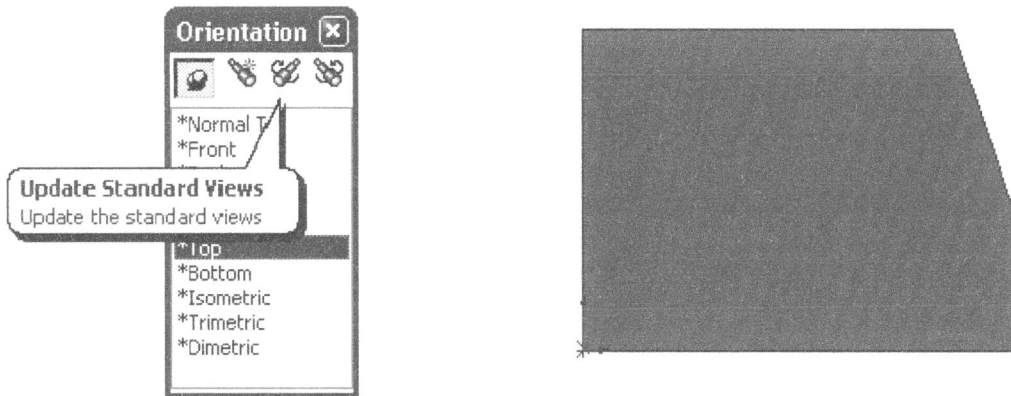

In the SolidWorks dialog box, click **Yes** to confirm the update.

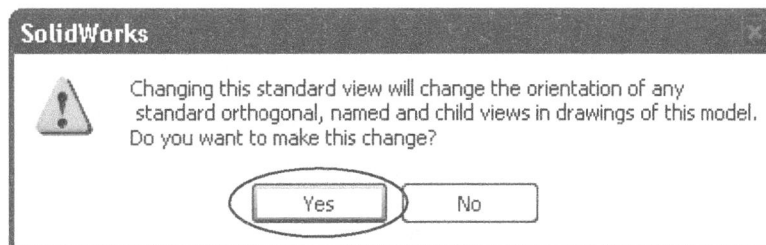

To see what this did, in the **Orientation** dialog box, double click on **Front** to change the orientation to the Front view. You can easily see that the standard views have been updated.

In the **Orientation** dialog box, double click on **Trimetric** to change the orientation to the Trimetric view.

Reset the Standard Views

To change things back to the original standard views, in the **Orientation** dialog box, click the **Reset Standard Views** button.

In the **SolidWorks** dialog box, click **Yes** to confirm the update.

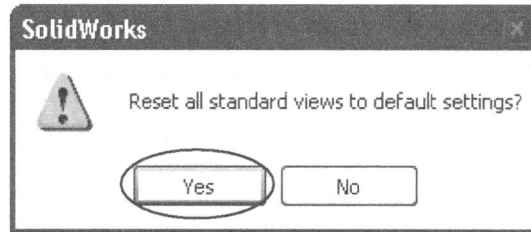

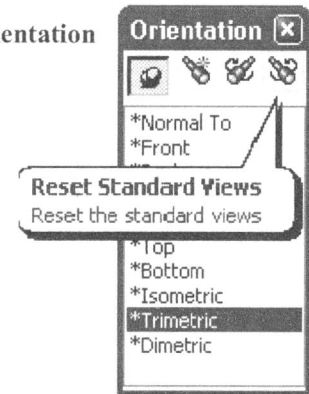

In the **Orientation** dialog box, double click on **Trimetric** to change the orientation back to the original Trimetric view.

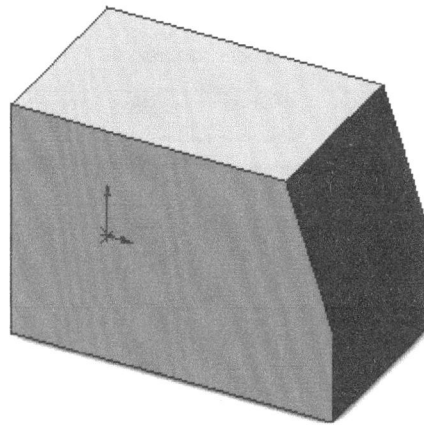

Add New Named View

You can use the Rotate, Zoom, and Pan commands to change the orientation of the model and save any orientation as a named view. For this example, you will create a new view using the angled face.

To do this, select the angled face and, in the **Orientation** dialog box, double click **Normal To**.

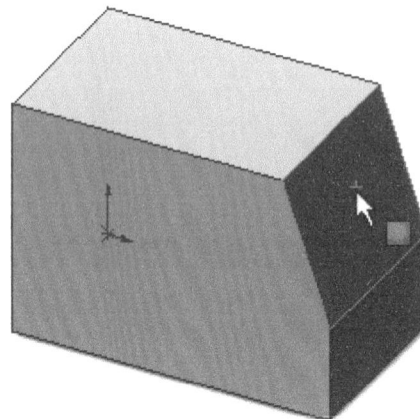

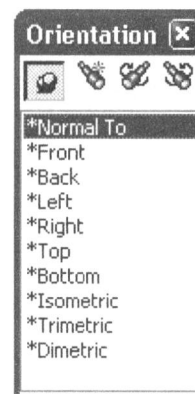

With the **Orientation** dialog box open, click the **New View** button.

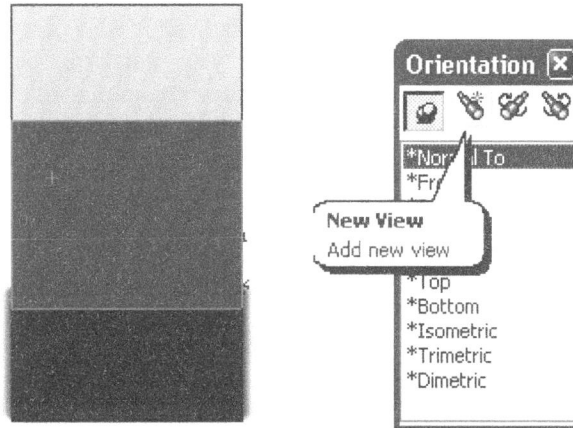

Type '**Face**' in the dialog box, and then click **OK**.

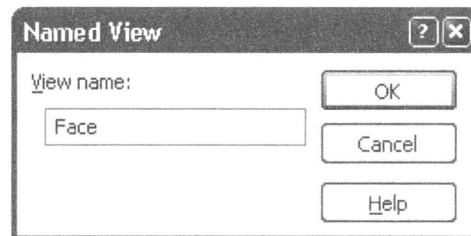

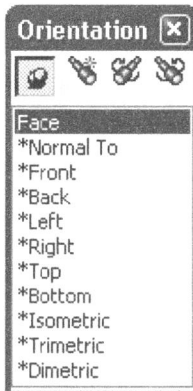

The name appears at the top of the list of views. There is no rename command. To rename the view, you must delete it and add a new named view. To delete a named view, select the name and press **Delete**.

But just for fun, make this new view the Top view. Then double click on the different standard views to check it out. Don't forget to reset the standard views when you are finished playing.

In the **Orientation** dialog box, double click on **Trimetric** to change the orientation back to the original Trimetric view.

To display the new view, simply double click on the view name, **Face**.

To close the **Orientation** dialog box, click the **Close** button in the upper right hand corner of the dialog box.

If you ever want to switch to a different view of the model and the **Orientation** dialog box is not visible, just press the **Space Bar**. Then, in the **Orientation** dialog box, double click the desired view. Another method is to just use the View list in the lower left corner of the graphics area.

Note that since the **Orientation** dialog box is pinned, the dialog box will reappear at the same pinned location each time. I prefer to keep this dialog box unpinned so it will appear at my cursor location and close itself when I pick a view, saving me mouse motion and an extra click. You can also close this dialog box just by clicking in the graphics area when the dialog box is unpinned.

Viewports

Now that you are familiar with different orientations of your part, it is time to take it to the next level. In addition to allowing you to view your part from any standard or custom view, SolidWorks also allows you to see your part in more than one view at the same time. Selecting an item in one view selects it in all views. You can view models through one, two, or four viewports. The options you picked when installing SolidWorks determine which views are shown. Note that you can change this at any time.

Pull down the "Tools" menu and pick **Options**.

In the **System Options** dialog box, on the **System Options** tab, pick **Display/Selection**.

At the bottom of the dialog box, set **Projection type for four view viewport** to **Third Angle** and click **OK**. **Third Angle** shows the Front, Right, Top, and Trimetric views. **First Angle** shows the Front, Left, Top, and Trimetric views.

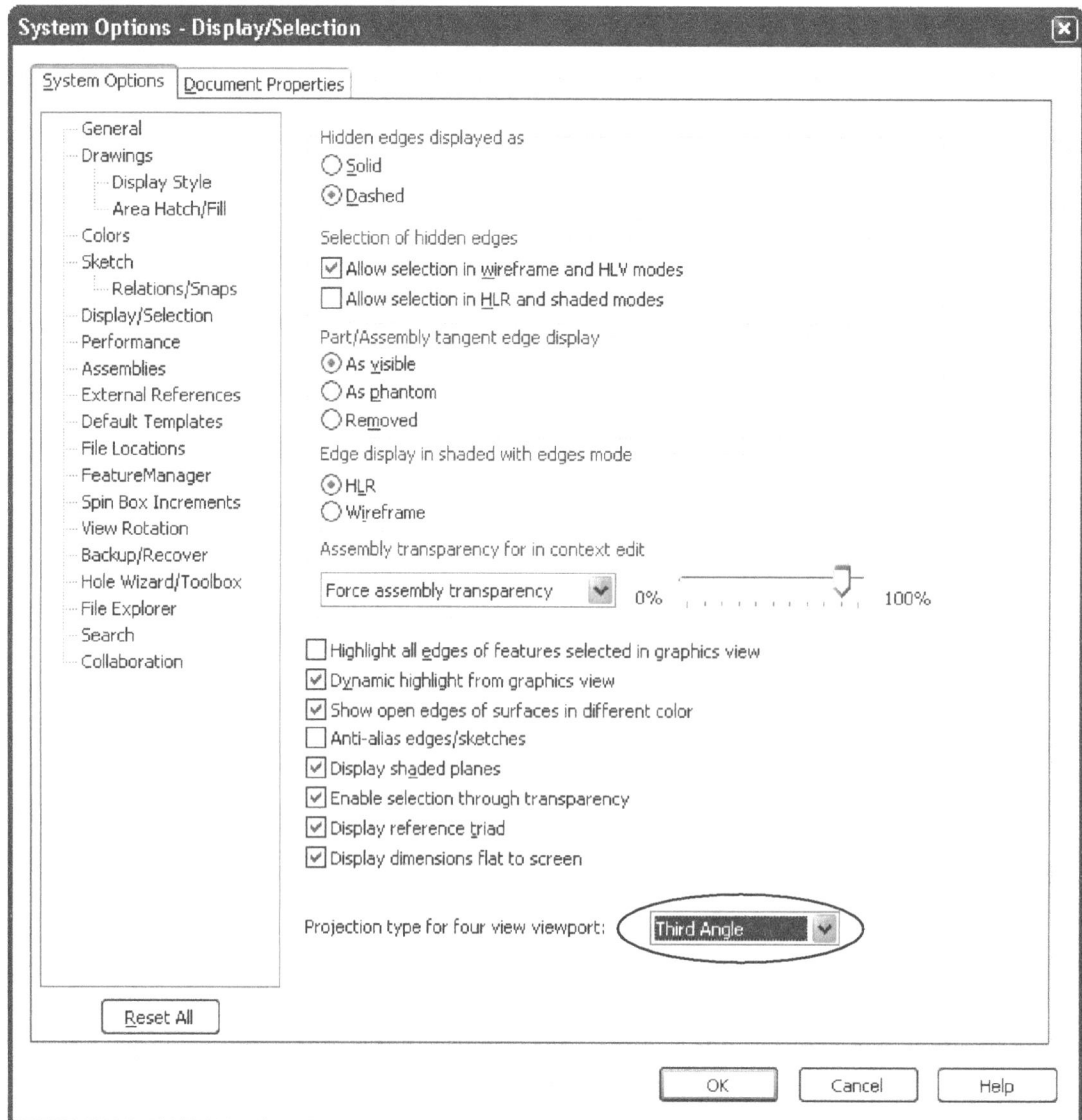

Two View – Horizontal

Pull down the "Window" menu and pick **Viewport – Two View - Horizontal** to display the front and top views in the viewport. This command can also be accessed from the View pop-up menu in the lower left corner of each viewport. The View pop-up menu displays the current orientation and is a different color in the active window than the pop-up menus in the other viewports. If the displayed views aren't the ones you wanted to see, you can change them. Just pick any view from the View pop-up menu of each viewport or rotate the model to a custom orientation.

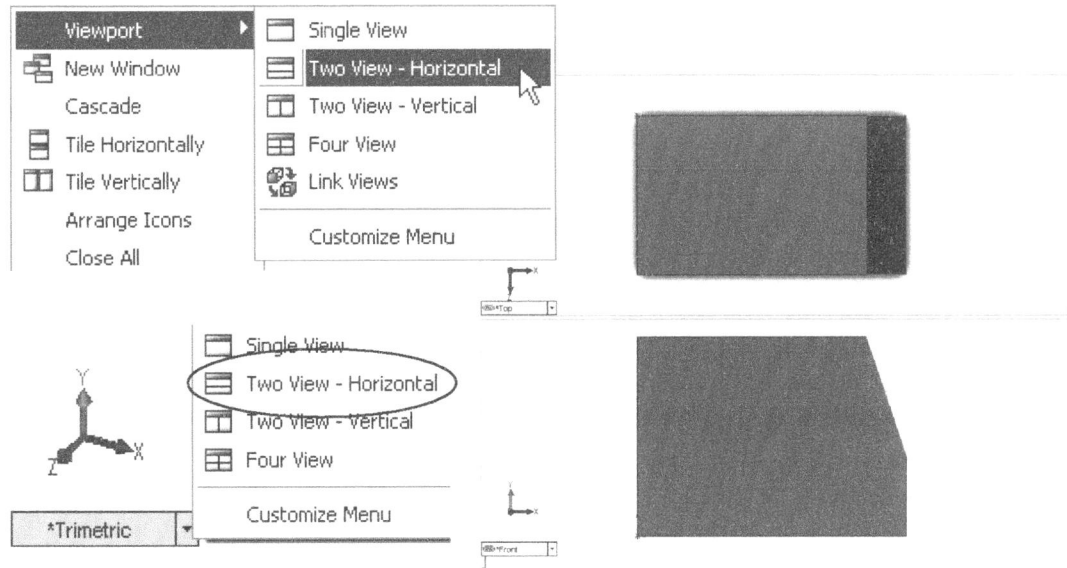

Two View – Vertical

Pull down the "Window" menu and pick **Viewport – Two View - Vertical** to display the front and right views in the viewport. This command can also be accessed from the View pop-up menu in the lower left corner of each viewport.

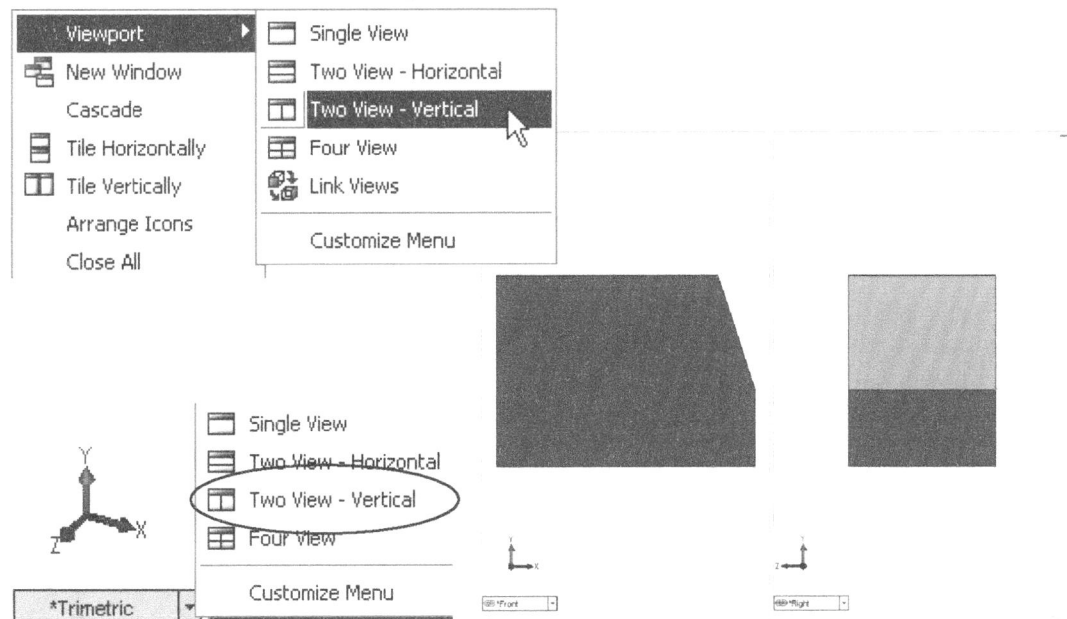

Single View

Pull down the "Window" menu and pick **Viewport – Single View** to display a viewport with a single view. This command can also be accessed from the View pop-up menu in the lower left corner of each viewport. When you switch from multiple viewports back to a single viewport, the model displays the orientation of the active viewport. If you then switch back to multiple viewports, the orientations are the defaults again.

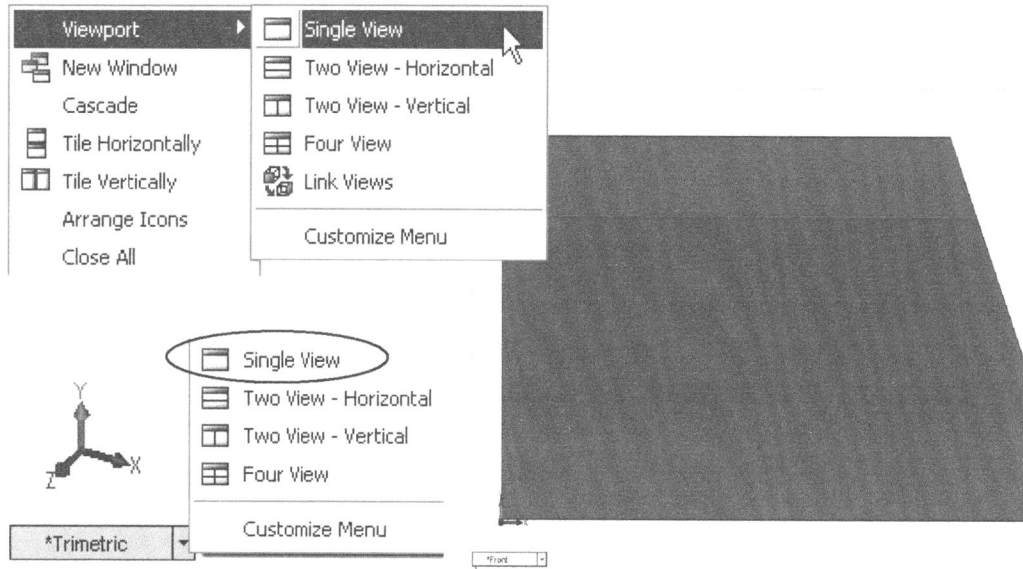

Four View

Now, to open four views of your part, pull down the "Window" menu and pick **Viewport – Four View**. This command can also be accessed from the View pop-up menu in the lower left corner of each viewport.

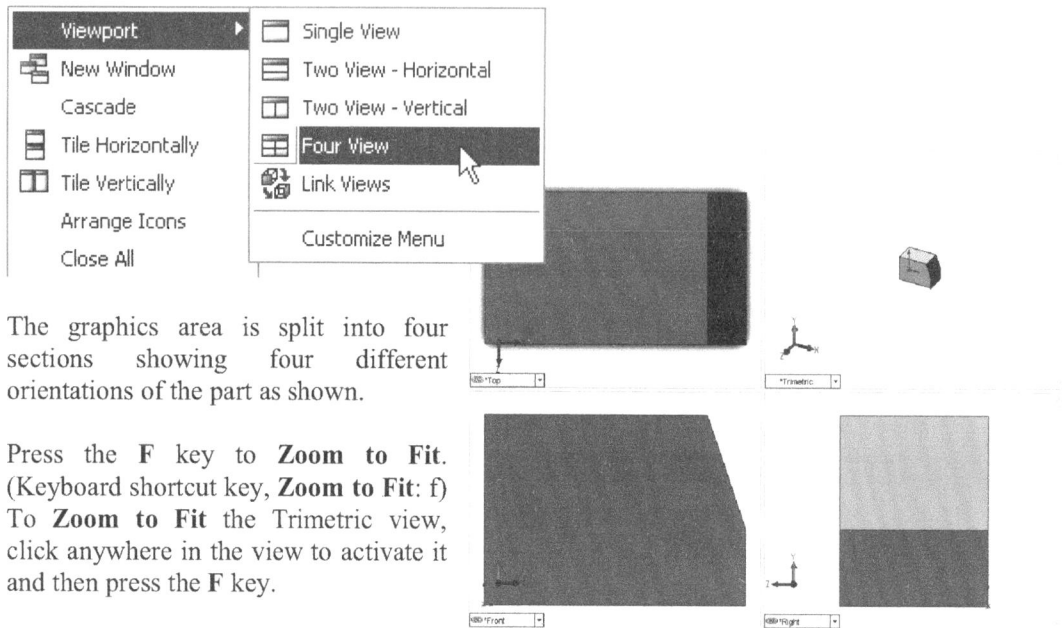

The graphics area is split into four sections showing four different orientations of the part as shown.

Press the **F** key to **Zoom to Fit**. (Keyboard shortcut key, **Zoom to Fit**: f) To **Zoom to Fit** the Trimetric view, click anywhere in the view to activate it and then press the **F** key.

Link Views

You can link the orthogonal views. Non-orthogonal views, such as Isometric, Trimetric, Dimetric, and custom orientations, can not be linked. If **Linked Views** is on and you rotate a view into orthogonal orientation, the view becomes linked. If you rotate a view out of its orthogonal orientation, it becomes a custom view and it is no longer linked.

By default the views are linked. The linked icon appears on the View pop-up menu when the viewports are linked. To toggle the link feature on or off, pull down the "Window" menu or open one of the pop-up menus and pick **Viewport – Link Views** to toggle **Link Views**.

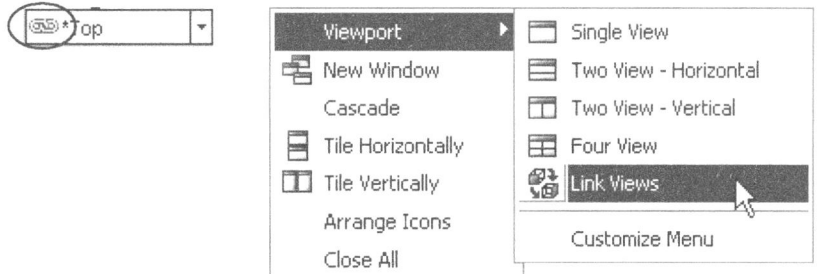

Zooming and panning are linked for all orthogonal views. For example, with **Link Views** active, hold down the middle mouse button or mouse wheel and drag to **Pan** the Front view, as shown.

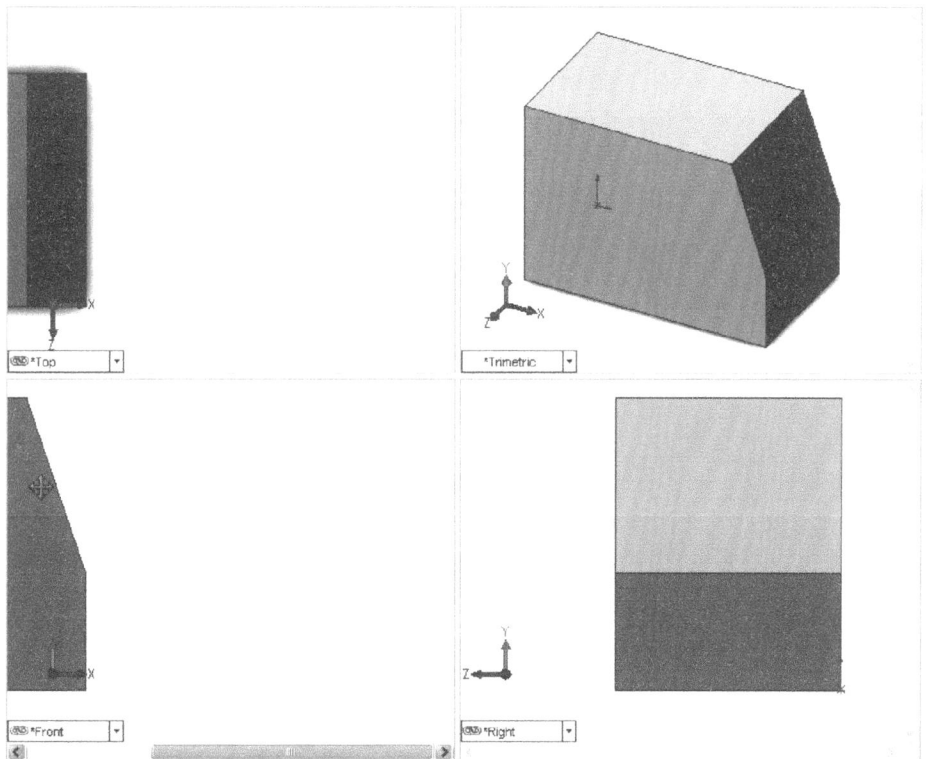

The Top and Right views move with the Front view.

Press the **F** key to **Zoom to Fit** all the linked views.

Split Controls

You don't have to use the pull down menu to split the graphics area. An easier way to display multiple views of the same document is to use the window split controls. You can use split controls to change the size and number of the panes.

When you place the cursor over the split bars which separate the different views, the cursor changes as shown.

Drag the horizontal and vertical split bars to change the size of the viewports as shown.

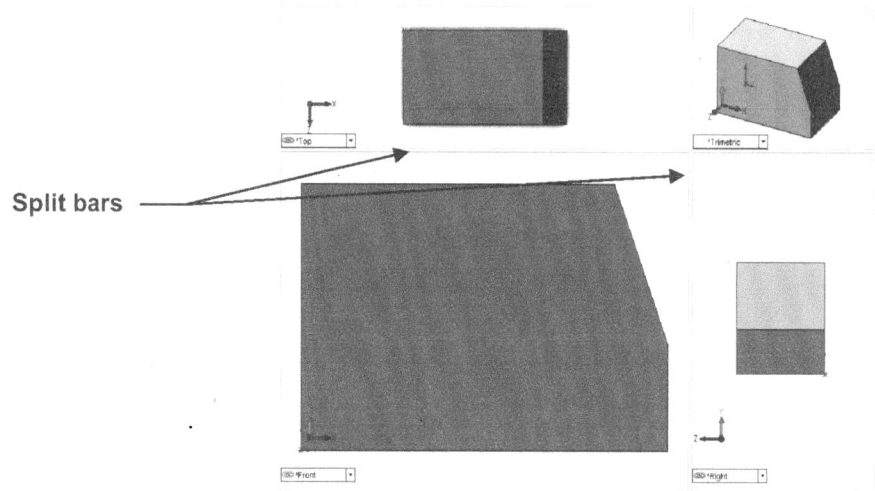

Split bars

Now, drag the vertical split bar to the right edge of the graphics area, to change from four views to two views. Note that the dragging to the left edge will produce the same result.

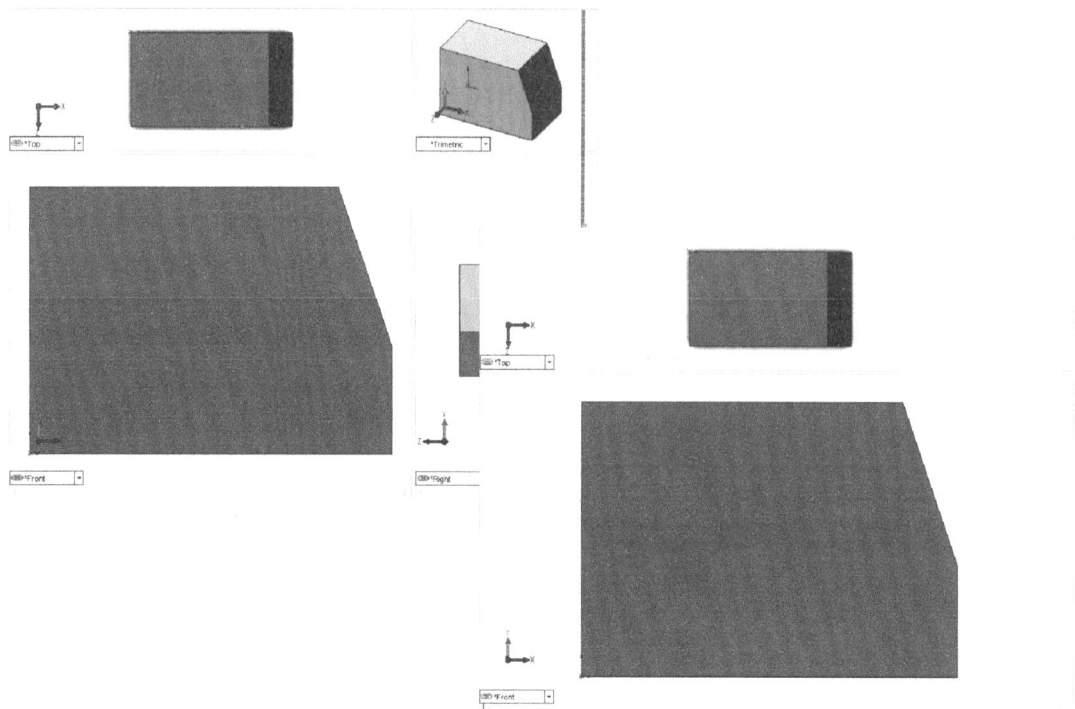

Now, drag the horizontal split bar to the top edge of the graphics area, to change from two views to a single view.

Move the cursor to the left end of the horizontal scroll bar at the bottom of the SolidWorks window, not the edge of the viewport but the end of the scroll bar. When the cursor is over the split control, the cursor will change as shown.

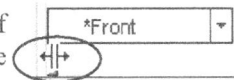

Drag the split control to the right and the graphics area will be split to any size that you want. You can drag the split bar at any time to change the size of the viewports.

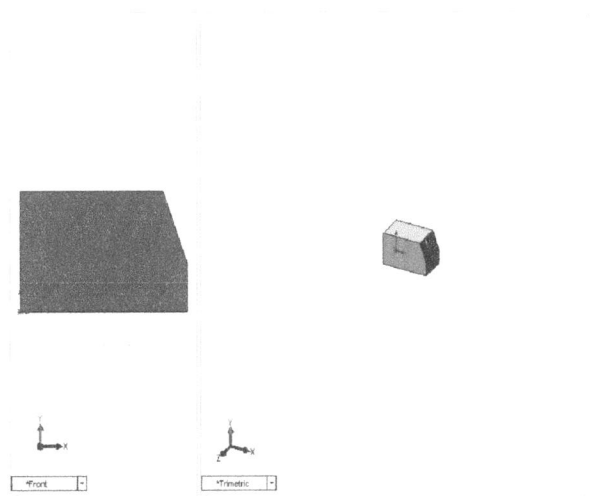

Now, move the cursor to the top of the vertical scroll bar at the right side of the SolidWorks window. This time, when the cursor is over the split control, and the cursor changes, double click on the split control.

The screen will be split in half vertically. Any hot keys that you press will only affect the active window. To change the active viewport, click anywhere inside the viewport that you want active. The View pop-up menu in the bottom left corner of the viewport will be solid, whereas the other View boxes are transparent. The same result occurs using the "Window" pull down menu commands. The difference is that default views are already defined for you, rather than having to manually set the view orientation of each view.

In the bottom left corner of the upper left viewport, change the View orientation by clicking the View pop-up menu and picking **Top**, or click in the upper left viewport and press **Ctrl+5**.

In the bottom left corner of the bottom right viewport, change the View orientation by clicking the View pop-up menu and picking **Right**, or click in the bottom right viewport and press **Ctrl+4**.

In the bottom left corner of the bottom left viewport, change the View orientation by clicking the View pop-up menu and picking **Front**, or click in the bottom left viewport and press **Ctrl+1**.

Press the **F** key on the keyboard to **Zoom to Fit**. The three linked views update. Click in the top right viewport and press the **F** key on the keyboard to **Zoom to Fit**.

If you move the cursor over the part and highlight any edge or entity, the corresponding edge highlights in all of the viewports. There is only one model you are working on. You are just looking at it four different ways simultaneously.

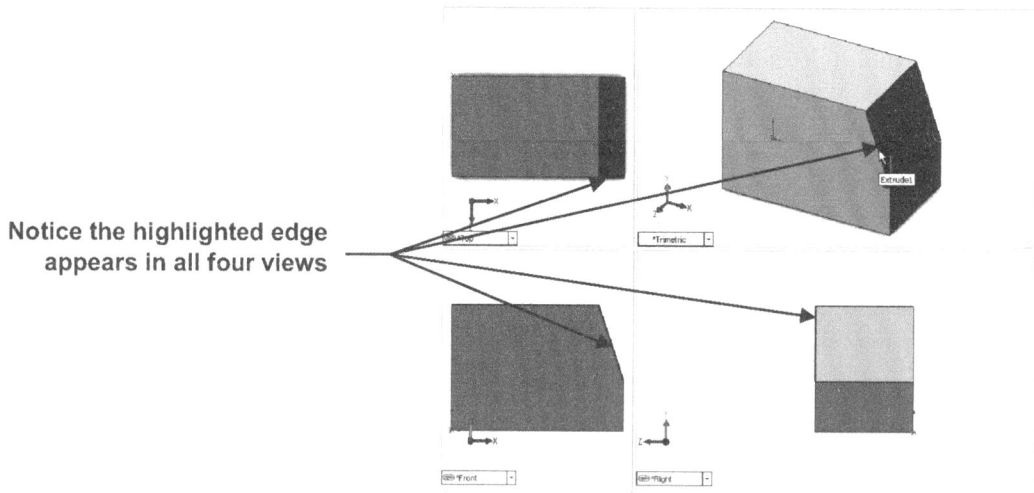

Notice the highlighted edge appears in all four views

Multiple Windows of the Same Document

An alternative to multiple viewports is to open multiple windows at the same time. This allows more freedom and flexibility in arranging and sizing multiple views of your part. You can open as many windows of the same document as you like, and you can close any of these open windows at any time. There is still only one model and all windows are updated simultaneously.

To understand this better, pull down the "Window" menu and pick **New Window**. Notice that there is just one viewport and the title bar (upper left corner of your screen) now shows the filename and an instance number.

SolidWorks - [Part1 *:2]

Pull down the "Window" menu again and pick **New Window**. Notice that the title bar now shows the filename and another instance number. Each of the document windows has its own FeatureManager design tree.

Right now, you have three windows of the same document open. By default, you only see one at a time. Pull down the "Window" menu to see your open files. The active file has a check mark in front of it.

Activate the first instance by picking it from the list as shown.

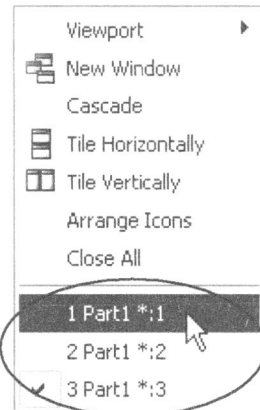

In the upper right hand corner of the SolidWorks window are two sets of buttons. Click the **Restore Down** button as shown to reduce the size of the open windows. You can now move the windows around by dragging the desired window by its title bar. You can also resize each window by dragging a corner or border.

Resize the first instance window to take up about two-thirds of the SolidWorks window as shown. Next, pull down the "Window" menu and pick the second instance, **Part1 *:2**. Then, resize the second instance window and move it to the top right corner of the SolidWorks window as shown. Finally, pull down the "Window" menu and pick the third instance, **Part1 *:3**. Then, resize the third instance window and move it to the bottom right corner of the SolidWorks window as shown.

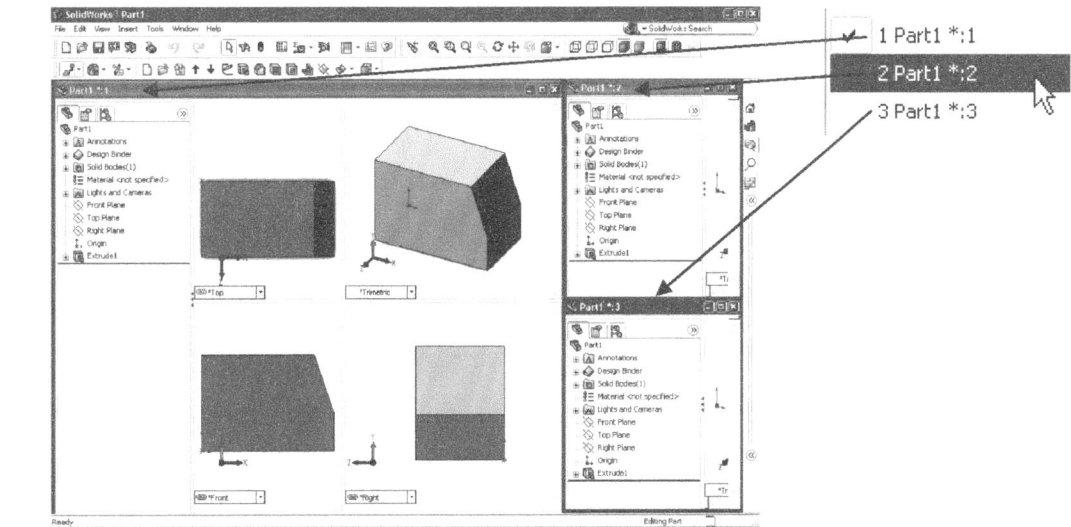

In the third instance window, click the **Hide FeatureManager Tree Area** bar as shown so that you will see only the document viewport. Do the same for the second instance window.

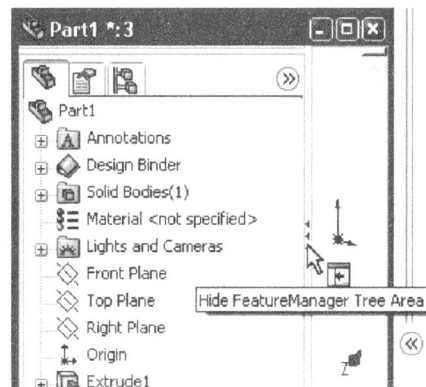

Now, in the first instance window, **Part1 *:1**, double click on the vertical split bar to return the graphics area to two viewports. Then, double click on the horizontal split bar returning to a single viewport.

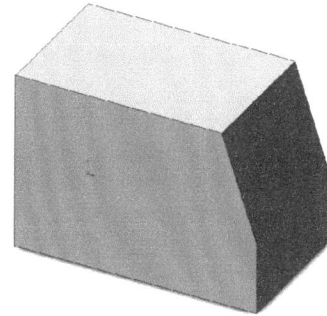

The active viewport is the viewport that remains. The other viewports are now closed.

In the FeatureManager design tree, right click on **Extrude1** and pick **Edit Feature**.

In the **Extrude** PropertyManager, under **Direction 1**, set the **Depth** to '**400**', and click the green check mark button at the top of the **Extrude** PropertyManager to accept the changes.

Press the **F** key to **Zoom to Fit**. If this doesn't work properly, don't get upset. Just use the **Zoom to Area** and **Pan** commands to fit the model on the window.

This is an interesting way to view a very long part. Rather than split the screen to four viewports, we are using three windows. You can see the entire part in the first instance window. You can zoom in the back end of the part in the second instance window and zoom in to the front end of the part in the third instance window as shown.

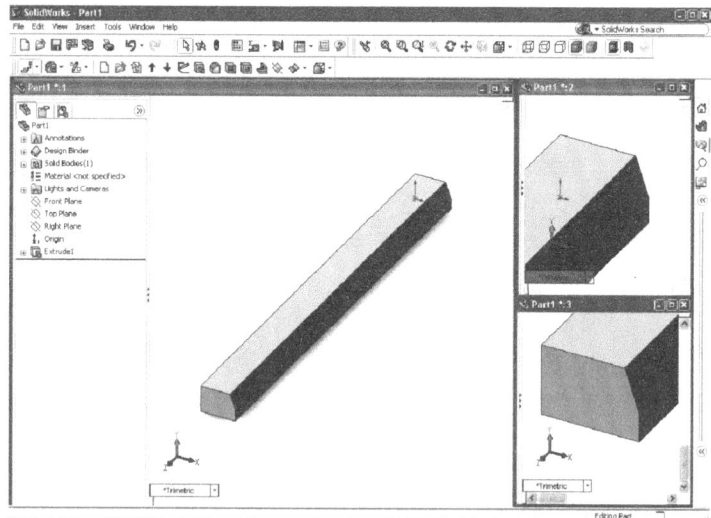

You can zoom or rotate the model in each of the windows independently. The **Linked Views** feature is not applicable. Any changes that you make to the model are reflected simultaneously in all of the windows.

To close an open window of the document, just click the **Close** button in the upper right hand corner of the document window. Not the main SolidWorks window.

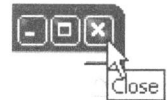

For now, pull down the "File" menu and pick **Close** to close the file completely.

SolidWorks will ask you to whether you want to save the document or not. You don't have to save the sample file for this chapter. You will not use it again.

The next chapter will show you even more about using multiple document windows.

Chapter 9

Multiple Document Interface

The Multiple Document Interface of SolidWorks allows multiple part, assembly, and drawing document windows to be open all at the same time. The File Explorer lets you preview and open multiple document files making it easier to tell you have the correct files.

The "Window" pull down menu in SolidWorks is a great way to manage all your open SolidWorks files. Besides allowing you to customize the display of multiple open documents, a list of the windows currently open is displayed, providing you quick access to each of the open windows.

Open Document Preview

Every part, assembly, and drawing file is referred to as a document, and each document is displayed in a separate window. There are several ways you can view these documents.

When you open a SolidWorks document, you have an option to display a preview of the part, assembly, or drawing document in the dialog box before actually opening it.

To do this, open up a previously created file by clicking the **Open** button from the "Standard" toolbar, or pull down the "File" menu and pick **Open**.

Browse to the default **<SolidWorks install directory>/Samples/tutorial/EDraw/claw** folder.

Make sure that the **Preview** check box is checked.

Click on **center.sldprt** to highlight it.

At the top of the dialog box, click on the **View Menu** button as shown on the next page and pick **Thumbnails** to view thumbnail images of each SolidWorks documents. The thumbnail preview is based on the view orientation of the model when the document was last saved.

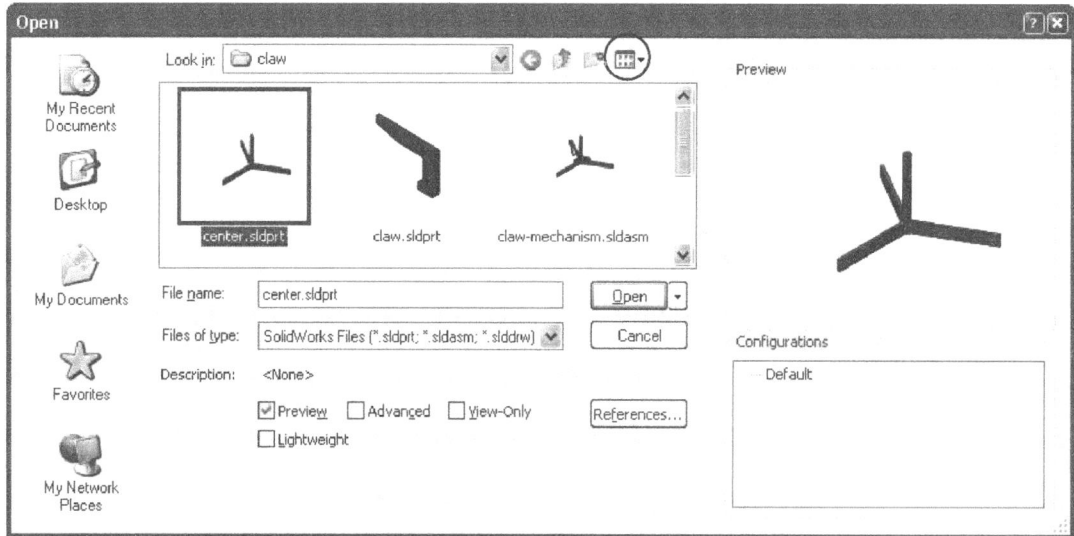

Click on the down arrow next to the **Open** button and pick **Open as Read-Only**. The **center.sldprt** file opens as a read-only file.

File Explorer

The **File Explorer** tab located in the Task Pane duplicates Windows Explorer from your local computer. You can specify what directories are displayed.

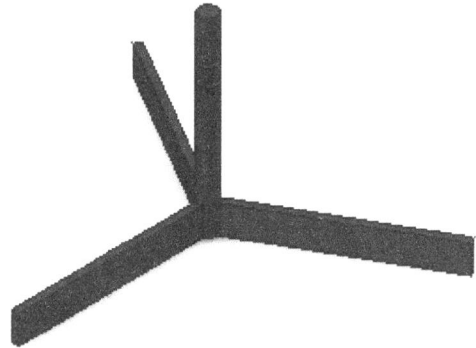

Pull down the "Tools" menu and pick **Options**.

In the **System Options** dialog box, on the **System Options** tab, click **File Explorer**.

For this chapter, in the **Show in File Explorer View** area, uncheck everything. Then, check **Samples** and click **OK**.

System Options - File Explorer

| System Options | Document Properties |

- General
- Drawings
  - Display Style
  - Area Hatch/Fill
- Colors
- Sketch
  - Relations/Snaps
- Display/Selection
- Performance
- Assemblies
- External References
- Default Templates
- File Locations
- FeatureManager
- Spin Box Increments
- View Rotation
- Backup/Recover
- Hole Wizard/Toolbox
- **File Explorer**
- Search
- Collaboration

Show in File Explorer view

- ☐ My Documents
- ☐ My Computer
- ☐ My Network Places
- ☐ Recent Documents
- ☐ Hidden referenced documents
- ☑ Samples

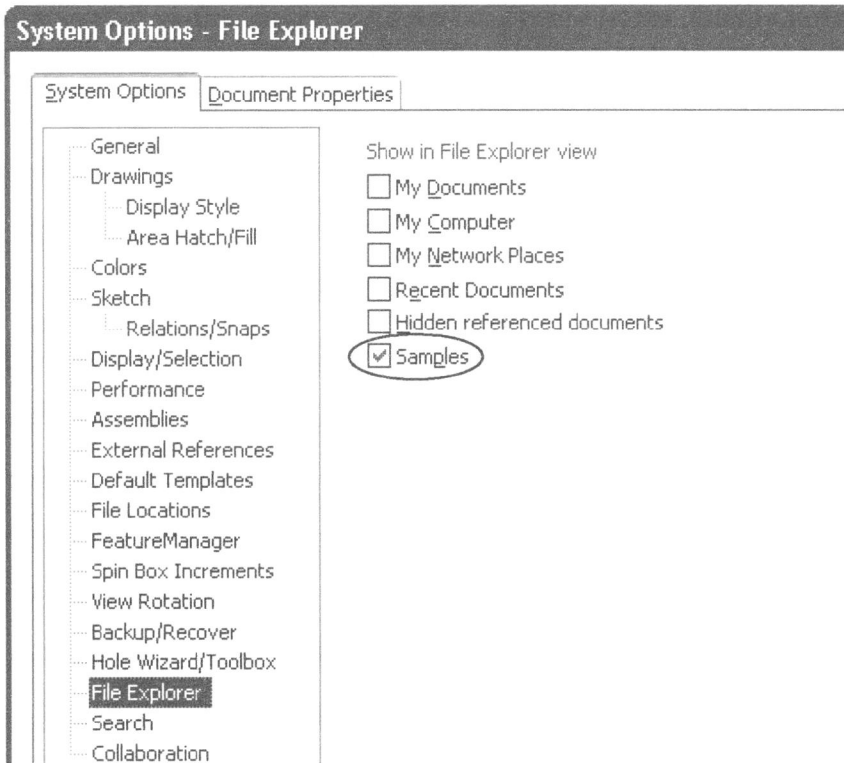

To show the Task Pane, pull down the "View" menu and make sure that **Task Pane** is checked. If not, check it.

In the Task Pane, click on the **File Explorer** tab to display the File Explorer.

Click the plus sign next to **Samples** to expand the folder. Click the plus sign next to the **tutorial** folder and then next to the **EDraw** folder and then next to the **claw** folder to expand the folders as shown.

Right click on **claw.sldprt** and pick **Open** from the menu. The document opens in the SolidWorks window. Because the windows are maximized, you only see the one file.

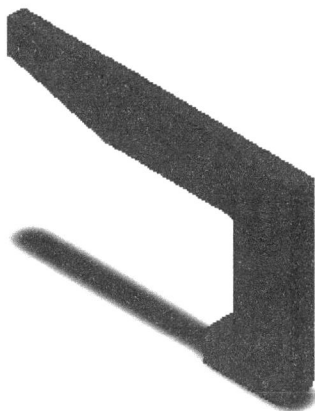

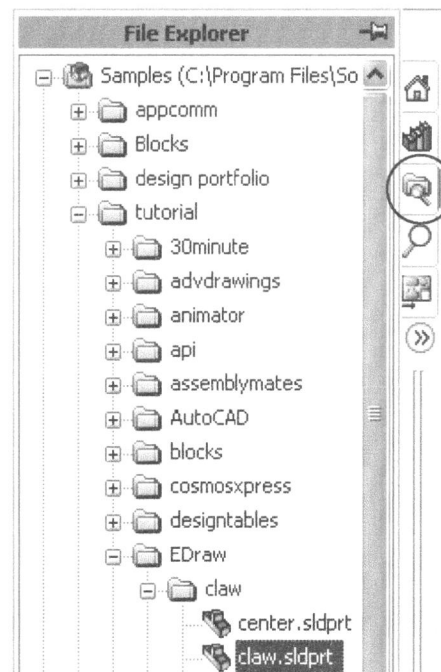

File Explorer

- Samples (C:\Program Files\So
  - appcomm
  - Blocks
  - design portfolio
  - tutorial
    - 30minute
    - advdrawings
    - animator
    - api
    - assemblymates
    - AutoCAD
    - blocks
    - cosmosxpress
    - designtables
    - EDraw
      - claw
        - center.sldprt
        - claw.sldprt

In the File Explorer, move the cursor over **claw-mechanism.sldasm**. A tooltip appears displaying the file name, a thumbnail preview, the path, date modified, and file size. If you can't see the file names, drag the left edge of the Task Pane to widen it.

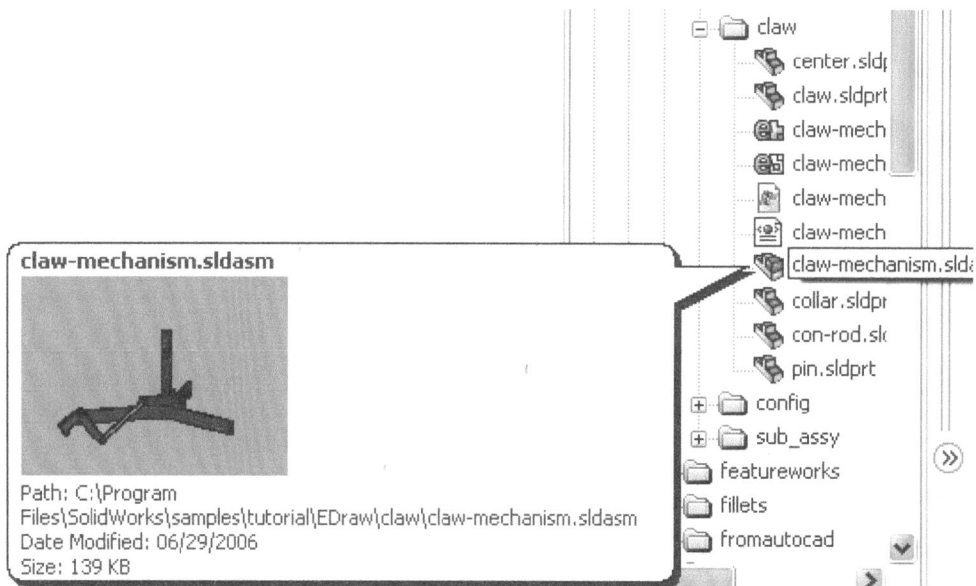

claw-mechanism.sldasm

Path: C:\Program
Files\SolidWorks\samples\tutorial\EDraw\claw\claw-mechanism.sldasm
Date Modified: 06/29/2006
Size: 139 KB

Press and hold the left mouse button on **claw-mechanism.sldasm** and drag the SolidWorks document to an empty area of the SolidWorks window. The assembly opens.

In the Task Pane, click on the **File Explorer** tab to display the File Explorer. In the upper right corner of the Task Pane, click the **Pin** button. Then, double click on **collar.sldprt**. This document opens in the SolidWorks window.

Under the **EDraw** folder, expand the **config** folder.

In the File Explorer, right click on **drw_crosshatch_solid.sldprt** and pick **Preview Window** from the menu.

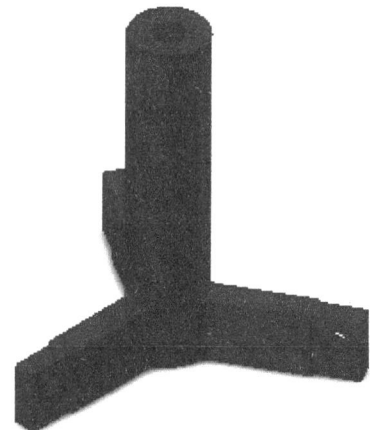

In the **Preview Window** dialog box, a **Document preview** is displayed. Below the preview, pull down the **Config** menu. A list of the document's configurations is shown. If you pick one of these configurations, it will display in the preview window. Otherwise, the message "Preview is not available" will be displayed.

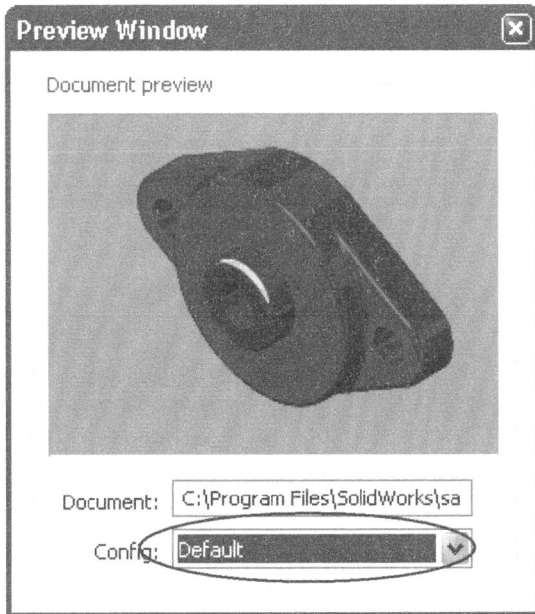

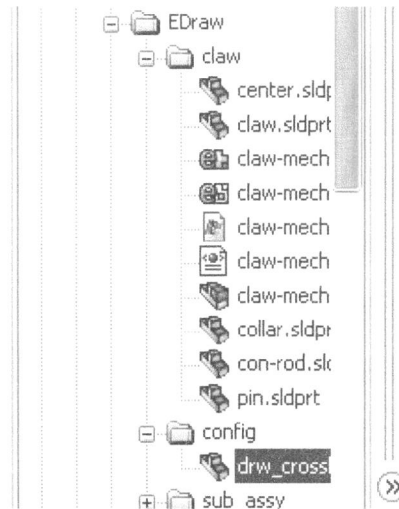

Press the **Escape** key to close this preview window.

Right click on **drw_crosshatch_solid.sldprt** and pick **Open to Configuration** and then pick **1.125** from the list. The document is opened and the **1.125** configuration is made active.

In the File Explorer, click the minus sign next to **Samples** to collapse the folder.

Click the plus sign next to **Open in SolidWorks** to expand the folder. A list of your open SolidWorks documents is shown. If not all your documents are shown, right click on **Open in SolidWorks** and pick **Refresh View**.

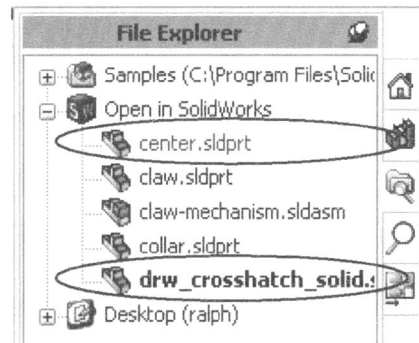

The **drw_crosshatch_solid.sldprt** file is bold because the document has been modified since it was last saved (the configuration was changed from the saved configuration). Since the **center.sldprt** file was opened as a read-only file, the file name is orange.

Right click on **drw_crosshatch_solid.sldprt** and pick **Save**. It is no longer bold, since the last save was of the current configuration.

Right click on **Open in SolidWorks** and pick **Show hidden referenced documents**. This option is also available by pulling down the "Tools" menu and picking **Options**. Then click on **File Explorer** and check **Hidden referenced documents**.

The icon is solid for an open SolidWorks file. The icon is transparent for the referenced assembly components that are in memory but not open in SolidWorks.

In the upper right corner of the Task Pane, click the **Pin** button.

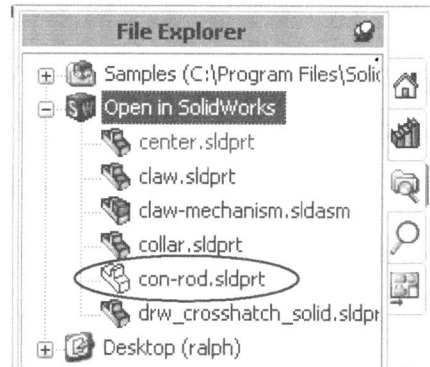

Customizing Different Views of SolidWorks Documents

Right now, you already have five files open. By default, you only see one document at a time. You can easily see which files are open by using the File Explorer.

To see what documents are open, pull down the "Window" menu. Here is the list of open windows. Remember, you can open multiple windows of the same open document in addition to opening different documents.

The active file has a check mark in front of it. To activate a different document, simply pick it from the list. Note that you can also use **Shift+Ctrl+Tab** to toggle forward and **Ctrl+Tab** to toggle backward through the open documents without pulling down the "Window" menu.

To organize the SolidWorks graphics area, you can change the size and location of the open document windows.

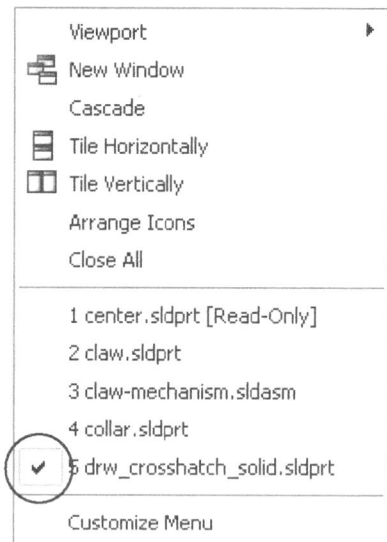

Cascade

Cascade is used to position multiple open windows in a cascade effect. When multiple windows are cascaded, the active window will be on top of the other windows.

Pull down the "Window" menu and pick **Cascade**.

You can now see all your open windows at the same time. At least a glimpse of each of them.

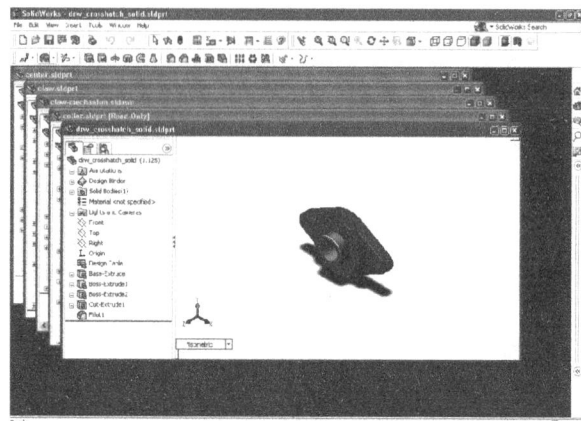

The title bar of the active document window, the one on top of the others, will be bright blue. The title bars of the other windows are transparent. All you have to do is click on any document window to pull it to the top and make it the active window.

To arrange the windows for better visibility, drag the desired window by its title bar. You can also resize the windows by dragging a corner or border.

Drag **drw_crosshatch_solid.sldprt** to the lower left of the SolidWorks Window and **collar.sldprt** to the lower right. Then resize them as shown so they do not over lap one another.

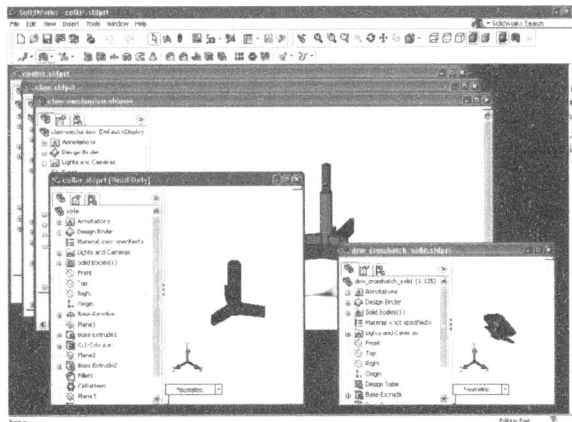

Each document window works the same as if you were working on it in a single viewport. Don't forget that you can also split each window separately into multiple viewports using the "Window" pull down menu or the View pop-up menu in the lower left corner of the viewport.

Click **Minimize** in the upper right corner of the **drw_crosshatch_solid.sldprt** document border.

A title bar, as shown, appears in the lower part of the SolidWorks window. If the title bar is not visible, it may be behind another open document.

Click **Restore Up** in the **drw_crosshatch_solid.sldprt** title bar to return the window back to its previous size. By the way, when a window is minimized, you can still drag the title bar around to move it to a more convenient screen location.

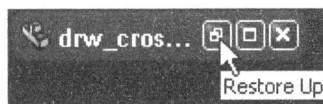

Click **Close** in the upper right corner of the **drw_crosshatch_solid.sldprt** document border to close this window.

Click **Close** in the upper right corner of the **collar.sldprt** document border to close this window.

Tile Horizontally

Tile Horizontally may be used to tile multiple windows one above another, splitting the screen horizontally. This option is ideal for displaying wide multiple documents.

Pull down the "Window" menu and pick **Tile Horizontally**.

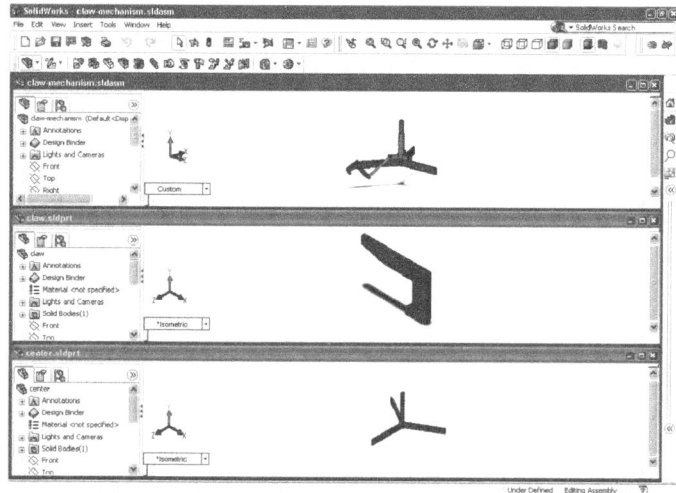

Tile Vertically

Tile Vertically may be used to tile multiple documents one beside another, splitting the screen vertically. This option is ideal for displaying vertically orientated documents.

When you have more than three windows open Tile Vertically and Tile Horizontally results in a grid pattern of the windows.

Also, you can close the FeatureManager design tree in any of the windows allowing more area for the model image.

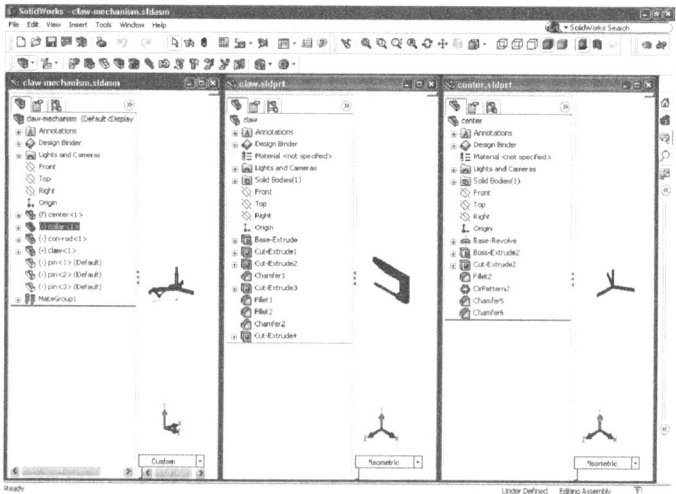

Pull down the "Window" menu and pick **Tile Vertically**.

When working with multiple files, you can easily save all of the open documents. Pull down the "File" menu and pick **Save All** to save all of the open documents.

Pull down the "Window" menu and pick **Close All** to close all of the open documents.

Chapter 10

Document Templates

When setting up SolidWorks, you should review and modify the default values based on your organization's needs. In years past, a common practice was to open up a new file, make changes to the default settings, and then save the file with a generic name like Blank. Instead of creating a new file, they would open up "Blank" to begin a new file and then use the 'Save as' command to save the new project with the correct name. Nowadays, this practice is not necessary because of the introduction of templates.

Part templates use a unique file extension, .prtdot. A template is essentially a blank design file that can be customized to contain specific settings. In other words, SolidWorks Part Templates are the starting settings for your SolidWorks .sldprt file. The template file stores the part settings and properties as well as the look and feel of the user interface. Even created geometry is saved with the template.

All these settings allow you to set up custom work environments. You can create multiple templates with specific customizations for different projects, using specific toolbars and menus for one project and a totally different set of menus and keyboard shortcuts for another project. These custom templates can contain special toolbars and menus to help you access certain functions more easily, but only when you need them. You can also get a head start on your part creation by saving a base feature predefined in your template. While this chapter is about part templates, the next chapter will cover drawing templates. Custom templates may be created for assembly documents as well.

Customize the Settings

At this point, we are going to create a part template by creating a new file and making the desired changes to the part document settings.

Press **Ctrl+N** to create a new **Part** file, or click the **New** button in the "Standard" toolbar.

In the FeatureManager design tree, right click on **Part1** and pick **Document Properties**. This option will appear if you click anywhere in the FeatureManager design tree with nothing selected.

On the **Document Properties** tab, change the **Dimensioning standard** to **ISO**. Then, check the **Dual dimensions display** check box and pick the **On bottom** radio button.

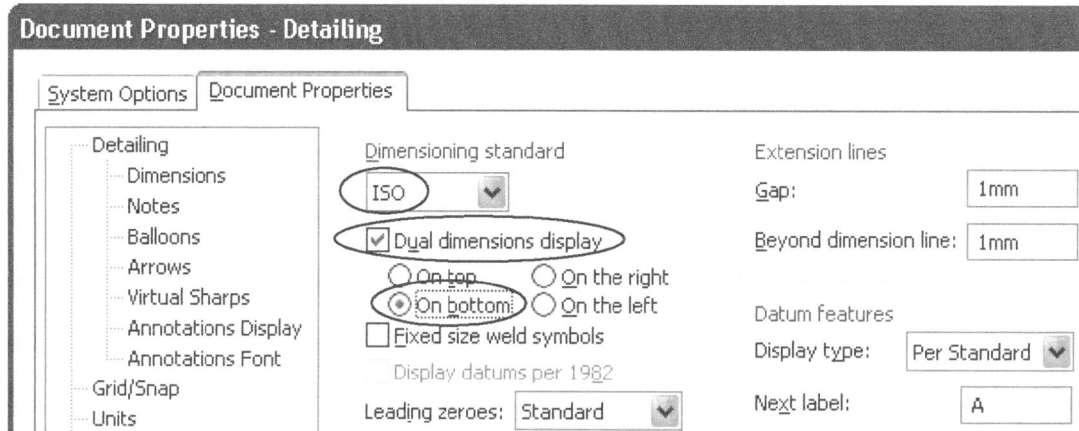

Next, click on **Units**, and change the **Unit system** to **MMGS (millimeter, gram, second)**. If you don't normally work with millimeters, that's OK. You can change the settings back later. For this example, millimeters will be used.

Under **Dual units**, make sure that **inches** is shown. Pick the **Fractions** radio button, check **Round to nearest fraction**, and change the **Denominator** to **16**.

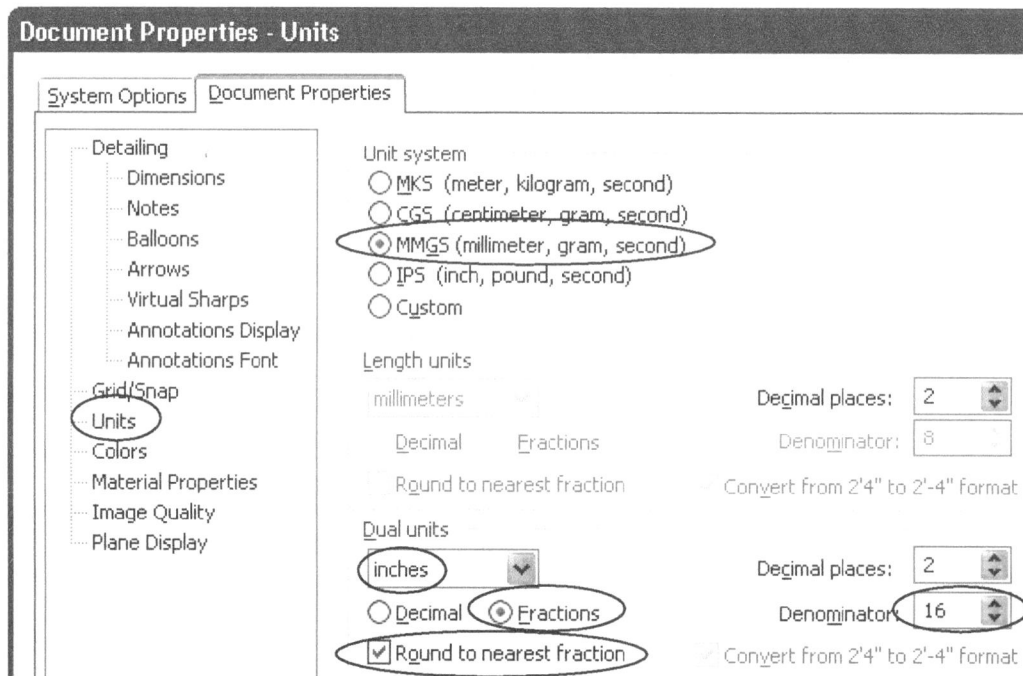

In the **Document Properties** dialog box, click **OK**. All of the settings that you change will be saved in your template and used for all parts created from this template. Note that these settings are unique to each document type: part, assembly, and drawing.

Create a Wood Block

Normally, a new file has no defined geometry features, meaning that you have to start from scratch each time. Using a predefined template with your basic geometry already defined, gives you a head start in modeling your part. By defining the initial features of a series of parts in your template, new parts can easily be created by building upon these initial features.

Create a base feature by clicking the **Explode Sketch** button in the control area of the CommandManager. Then, click the **Extruded Boss/Base** button from the toolbar, or pull down the "Insert" menu and pick **Boss/Base – Extrude**.

Select the **Top** plane when prompted to select a plane on which to sketch the feature cross-section.

Create a rectangle with the origin at the lower left corner using the **Rectangle** button in the CommandManager, or pull down the "Tools" menu and pick **Sketch Entities – Rectangle**.

Add the '50' and '150' dimensions as shown using the **Smart Dimension** button in the CommandManager, or pull down the "Tools" menu and pick **Dimensions – Smart**.

Exit the sketch by clicking the **Exit Sketch** button in the CommandManager or in the upper right corner of the graphics area.

In the **Extrude** PropertyManager, under **Direction 1**, set the **Depth** to '15'.

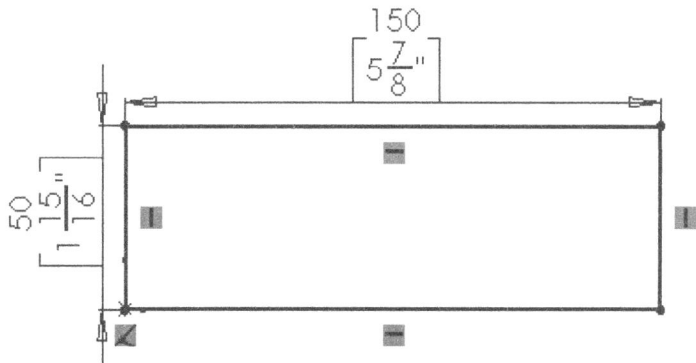

Click the green check mark button at the top of the **Extrude** PropertyManager.

Add Multiple Configurations

If parts created from a template will require multiple configurations, you can predefine those configurations in your template.

In the FeatureManager design tree, click on the **ConfigurationManager** tab.

Right click on **Part1 Configuration(s)** and pick **Add Configuration**.

Click in the **Configuration name** box and type '**With Holes**', and then click the green check mark button at the top of the **Add Configuration** PropertyManager.

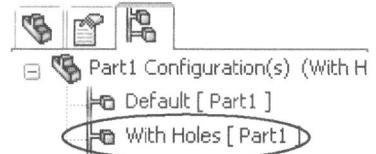

The **With Holes** configuration now appears in the configuration list and is the active configuration.

Create a hole by clicking the **Extruded Cut** button from CommandManager, or pull down the "Insert" menu and pick **Cut – Extrude**.

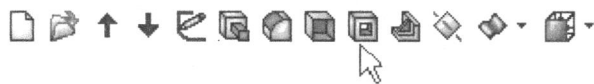

Pick the top of the part.

Add a **Ø9.50 [3/8"]** dimension to the circle as shown. Then, add the two **22.375 [7/8"]** dimensions.

To make your dimensions easier to read when looking at the Trimetric view, a pull down the "Tools" menu and pick **Options**.

In the **System Options** dialog box, pick **Display Selection** and check **Display dimensions flat to the screen**.

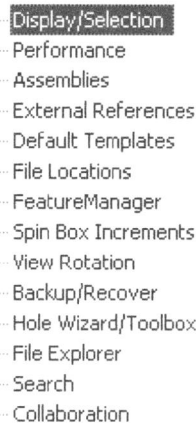

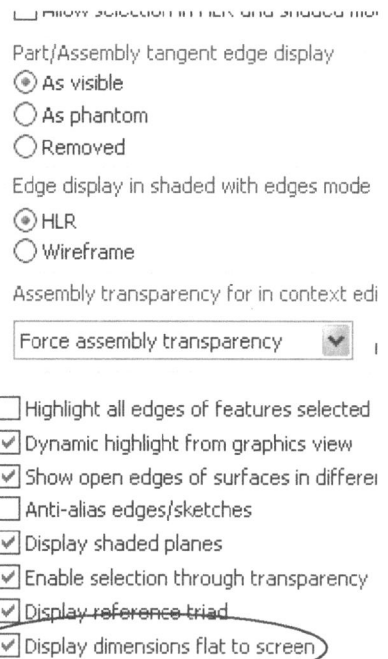

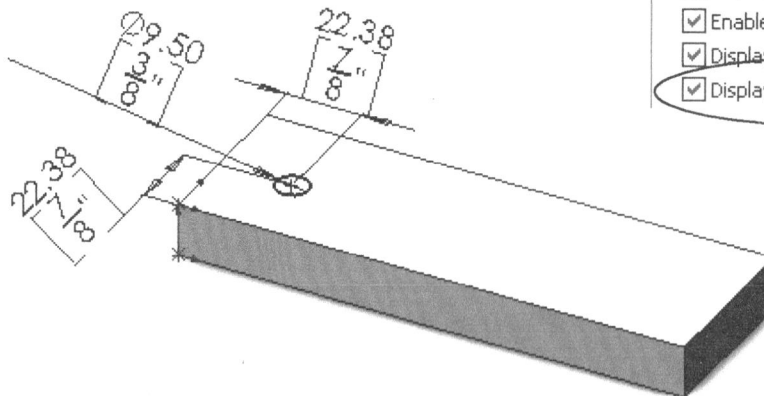

Display/Selection
Performance
Assemblies
External References
Default Templates
File Locations
FeatureManager
Spin Box Increments
View Rotation
Backup/Recover
Hole Wizard/Toolbox
File Explorer
Search
Collaboration

Part/Assembly tangent edge display
- As visible
- As phantom
- Removed

Edge display in shaded with edges mode
- HLR
- Wireframe

Assembly transparency for in context edi

Force assembly transparency

☐ Highlight all edges of features selected
☑ Dynamic highlight from graphics view
☑ Show open edges of surfaces in differei
☐ Anti-alias edges/sketches
☑ Display shaded planes
☑ Enable selection through transparency
☑ Display reference triad
☑ Display dimensions flat to screen

Add a Favorite

You can define favorite styles for dimensions and various annotations. This favorite setting will then be available whenever you create a part using this template.

To add a favorite, click on the **Ø9.50 [3/8"]** dimension to highlight it.

In the **Dimension** PropertyManager, change the **Primary Unit Precision** to **.1**.

Click the **Add or Update a Favorite** button. Note that all the changes to the settings in the **Dimension** PropertyManager are saved as one favorite. You can save many different groups of settings as favorites.

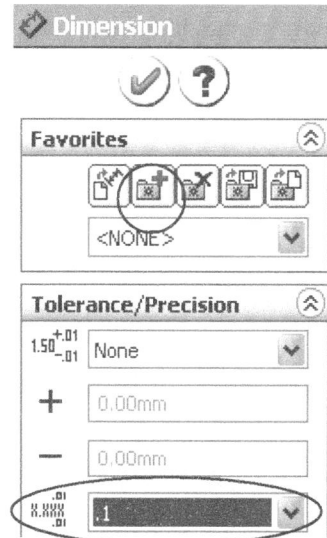

In the **Add or Update a Favorite** dialog box, in the **Enter a new name or choose an existing name** box, type '**One Decimal Place**' and click **OK**.

Click the green check mark button at the top of the **Dimension** PropertyManager. Note that the **Ø9.50 [3/8"]** dimension is now **Ø9.5 [3/8"]**.

To use the favorite, click on one of the **22.38 [7/8"]** dimensions in the graphics area.

In the **Dimension** PropertyManager, pull down the "Set a current Favorite" menu and pick **One Decimal Place**.

Click the green check mark button at the top of the **Dimension** PropertyManager.

Do the same thing for the other **22.38 [7/8"]** dimension.

Both dimensions are now **22.4 [7/8"]**.

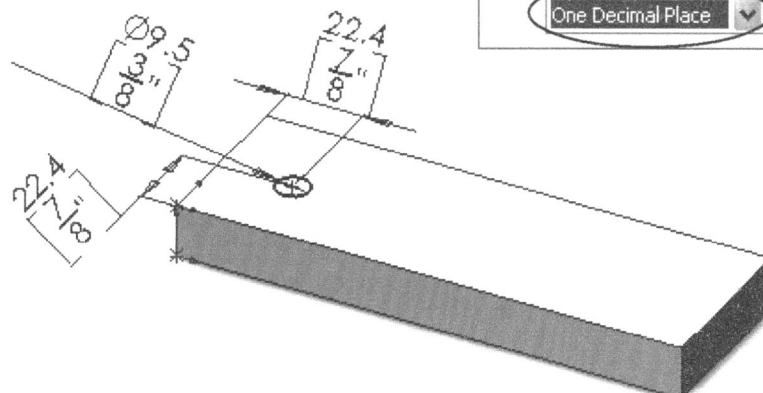

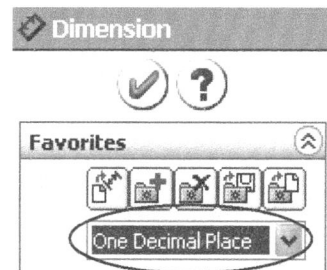

Exit the sketch by clicking the **Exit Sketch** button in the CommandManager or in the upper right corner of the graphics area.

In the **Cut-Extrude** PropertyManager, under **Direction 1**, pick **Through All**, and then click the green check mark button at the top of the **Cut-Extrude** PropertyManager.

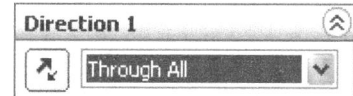

Direction 1	⊗
↗	Through All ▾

Customize the FeatureManager design tree

Take a moment and set the FeatureManager design tree options to the way that you want them so that they are also saved in your template.

Click on the **FeatureManager design tree** tab.

Then, right click on **Extrude1** and pick **Feature Properties**.

In the **Feature Properties** dialog box, type '**Wood Block**' for the **Name** and type '**Cedar**' for the **Description**.

Pick **All Configurations** from the pull down menu, and then click **OK**.

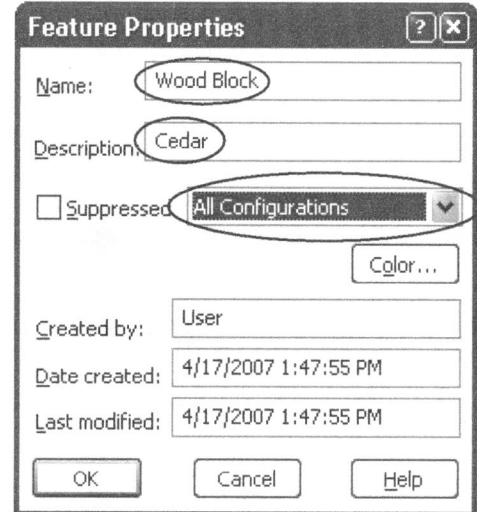

Feature Properties dialog box:

Name:	Wood Block
Description:	Cedar
☐ Suppressed	All Configurations ▾
	Color...
Created by:	User
Date created:	4/17/2007 1:47:55 PM
Last modified:	4/17/2007 1:47:55 PM
OK	Cancel	Help

In the FeatureManager design tree, right click on **Part1 (With Holes)** and pick **Tree Display – Show Feature Descriptions** to show the description of the feature. The result in this example is **Wood Block "Cedar"**.

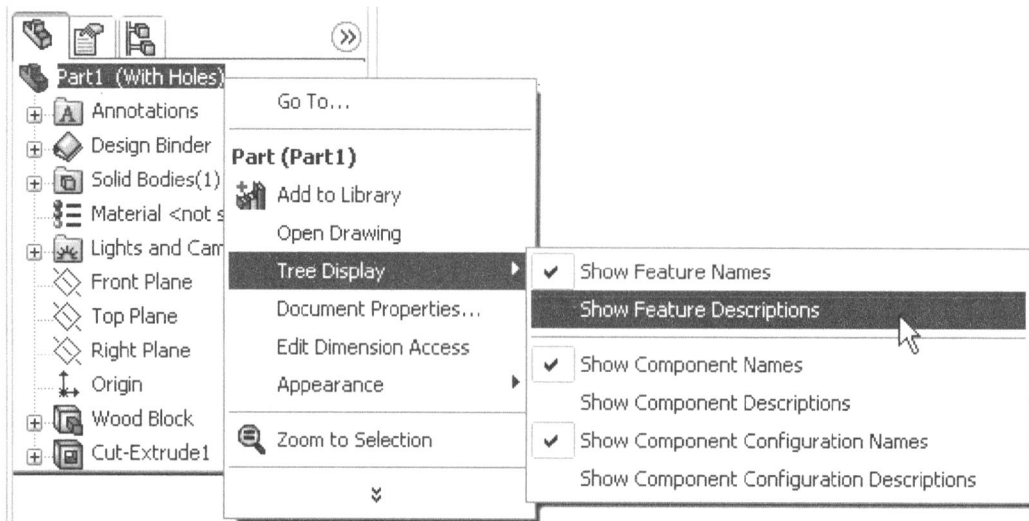

Next, pull down the "Tools" menu and pick **Options**.

On the **System Options** tab, click on **FeatureManager**.

Check **Name feature on creation** if you want the feature name in the FeatureManager design tree to automatically be in edit mode when a new feature is created. Click **OK**.

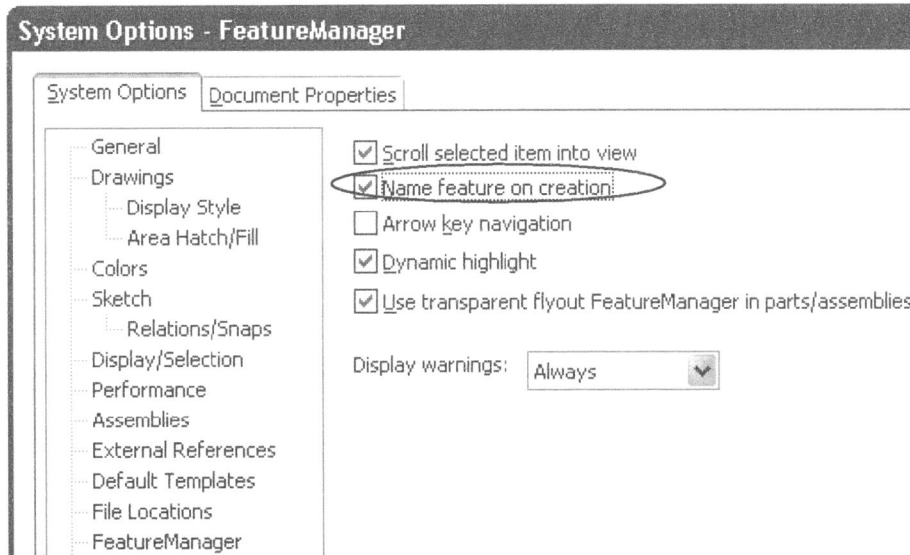

In the FeatureManager design tree, right click on the **Annotations** folder and pick **Show Feature Dimensions**.

Then, pull down the "Tools" menu and pick **Options**.

On the **System Options** tab, under **General**, check **Show dimension names**, and click **OK**.

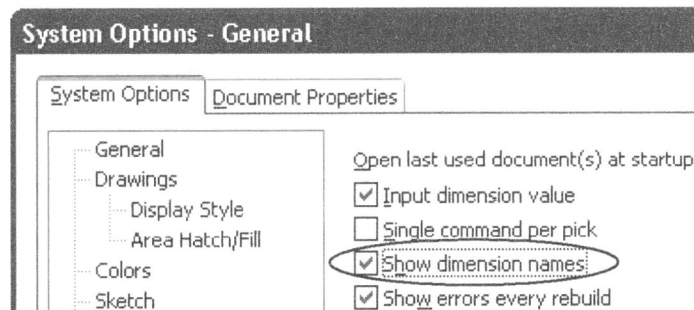

As you can see, the names are right on top of the dual dimensions.

To fix this, in the FeatureManager design tree, right click on **Part1** and pick **Document Properties**.

On the **Document Properties** tab, change the dual dimension display by picking the **On the right** radio button and click **OK**.

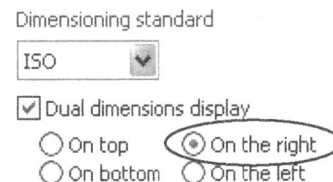

Right click on the **150 [5 7/8"]** dimension and pick **Properties** from the menu.

In the **Dimension Properties** dialog box, click in the **Name** box, type '**Length**', and click **OK**.

Right click on the **50 [1 15/16"]** dimension and pick **Properties**. In the **Dimension Properties** dialog box, click in the **Name** box, type '**Width**', and click **OK**.

Right click on the **15 [9/16"]** dimension and pick **Properties**. In the **Dimension Properties** dialog box, click in the **Name** box, type '**Height**', and click **OK**.

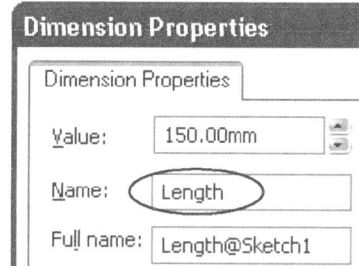

Drag the dimensions with the dimension names to arrange them so that you can easily see them.

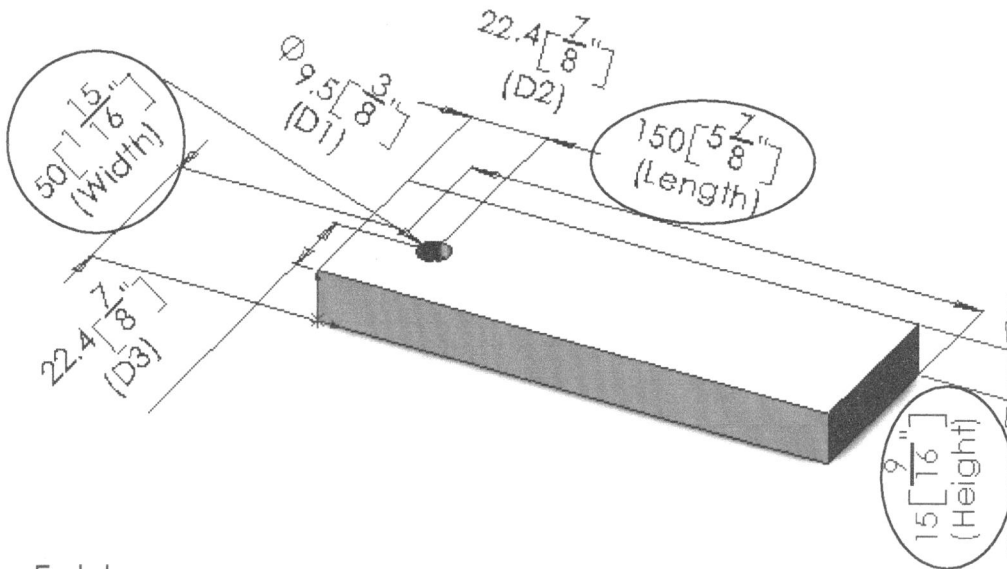

Add a Folder

In the FeatureManager design tree, right click on **Cut-Extrude1** and pick **Add to New Folder**.

Type '**Holes**' for the name of the folder. The feature, **Cut-Extrude1** is moved to this folder.

Adding folders to your FeatureManager design tree provides a convenient method of organizing the features of your model. Think about a part with over 100 holes. You could make a folder for each hole size and place the holes in the folders accordingly. Another nice feature is that when you click on a folder in the FeatureManager design tree, all of the model features in the folder highlight in the graphics area. Maybe your thing is fillets, instead of holes, the same concept applies.

Add a Comment

Like having a little notepad ties to each feature, you can add comments to remind you of what needs to be done. Here a comment is added to the Base feature to provide instruction for future use of the template model.

In the FeatureManager design tree, right click on **Wood Block "Cedar"** and click the down arrows (˅) at the bottom of the menu.

Then, pick **Comment – Add Comment**, as shown.

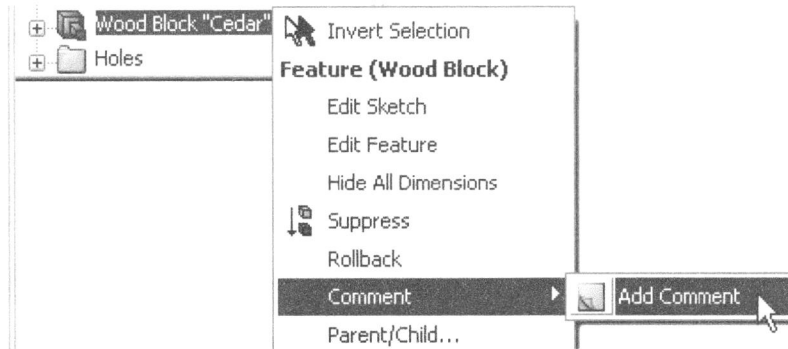

Click the **Date/Time Stamp** button to add the current date and time.

Then, in the message box, type '**Change these values to what you need**'.

Click the **Save and Close** button.

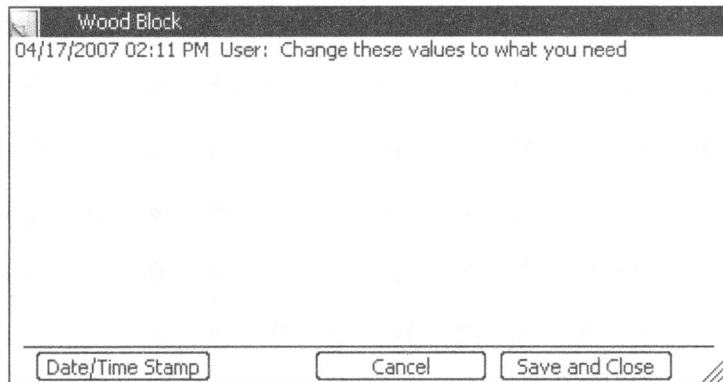

Now, when you place the cursor over the feature in the FeatureManager design tree, the note appears as a flyout.

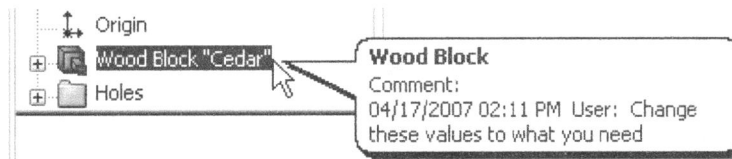

A better use for this feature is to add comments to remind you when a feature was revised and why it was revised. We all remember today why we made a change, but next year we won't have a clue why it was changed.

Split the FeatureManager design tree

When you are working with configurations or the design tree is just too long, you can split the tree area into two areas. This will allow you to view the FeatureManager design tree in one and the ConfigurationManager in the other. Or you can view two different segments of the design tree simultaneously.

To split the FeatureManager design tree, move the cursor to the top of the panel until it changes as shown below.

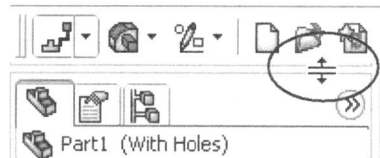

Press and hold the left mouse button as you drag the bar down below the last item in the panel. You can also double click on the split bar. Each instance of the panel displays a set of tabs.

In the upper panel, pick the **ConfigurationManager** tab. Move the split bar if necessary to allow more of the FeatureManager design tree to be visible. It probably is longer than the configuration list anyway. Also, at the time of writing, the keyboard shortcut **Shift+C** only collapses the FeatureManager design tree if it is in the lower panel.

Note that to close the split panel display, just double click the bar in between the panels, or drag the bar as shown up to its original position.

In the FeatureManager design tree, right click on **Material** and pick **Edit Material**.

In the **Materials Editor** PropertyManager, pick **Woods – Cedar**.

Click the green check mark button at the top of the **Materials Editor** PropertyManager to accept the settings.

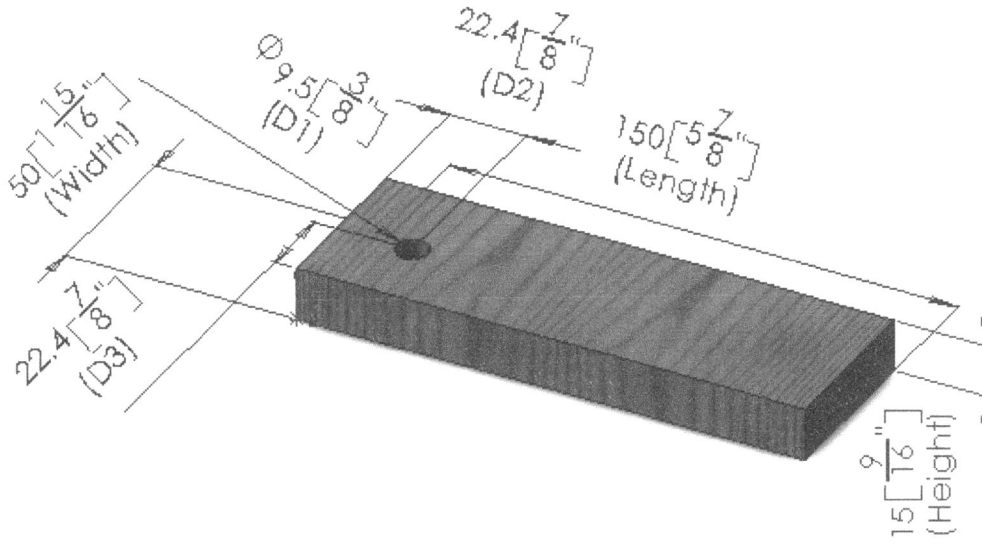

Create Custom Properties

Pull down the "File" menu and pick **Properties**.

In the **Summary Information** dialog box, on the **Summary** tab, click in the **Author** box and enter your name.

Then, click in the **Title** box and type '**Wood Block**'.

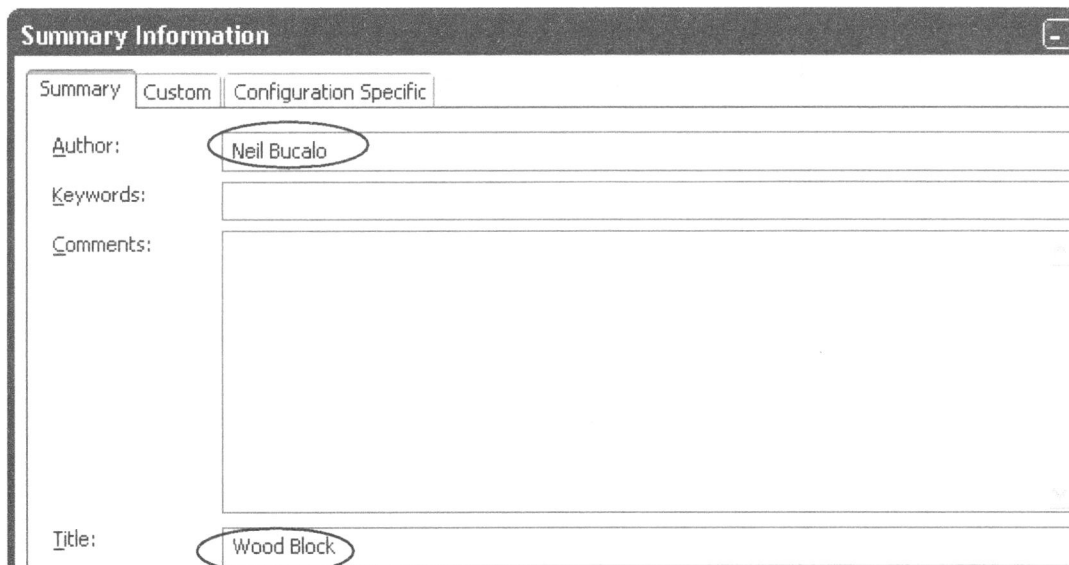

In the **Summary Information** dialog box, click on the **Custom** tab.

Click in the box under **Property Name**, pull down the list and pick **Description**. Keep **Type** set to **Text**.

Click in the **Value / Text Expression** box and type '**Wood Block**' and press **Enter**. Do not pull down the list, you must type this one.

In row 2, under **Property Name**, pick **DrawnBy** from the pull down list. Keep **Type** set to **Text**.

Click in the **Value / Text Expression** box and type your initials and press **Enter**.

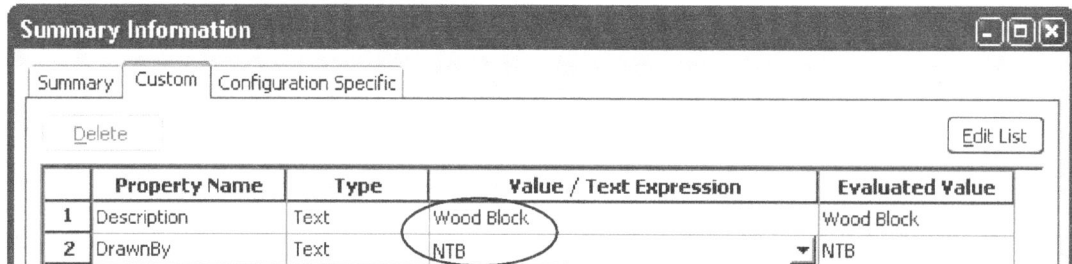

	Property Name	Type	Value / Text Expression	Evaluated Value
1	Description	Text	Wood Block	Wood Block
2	DrawnBy	Text	NTB	NTB

In the **Summary Information** dialog box, click **OK**.

Custom Mass Properties

Pull down the "Tools" menu and pick **Mass Properties**.

In the **Mass Properties** dialog box, click the **Options** button.

In the **Mass/Section Property Options** dialog box, under **Units**, click the **Use custom settings** radio button.

Then, change the **Decimal places** to **4**. Note that these custom settings apply only to values related to the **Mass Properties** and **Section Properties** dialog boxes. They do not override document properties for any other mass property functions.

Click **OK**. You can see that all the mass properties of the part in the **Mass Properties** dialog box are now shown with four decimal places.

Click the **Close** button.

Save As a Template

Pull down the "File" menu and pick **Save As**.

Change the **Save as type** to **Part Templates (*.prtdot)**.

Make sure that the **Save in** folder is the **<SolidWorks Install Directory>/templates** directory.

Type '**Cedar**' for the **File name** and click **Save**.

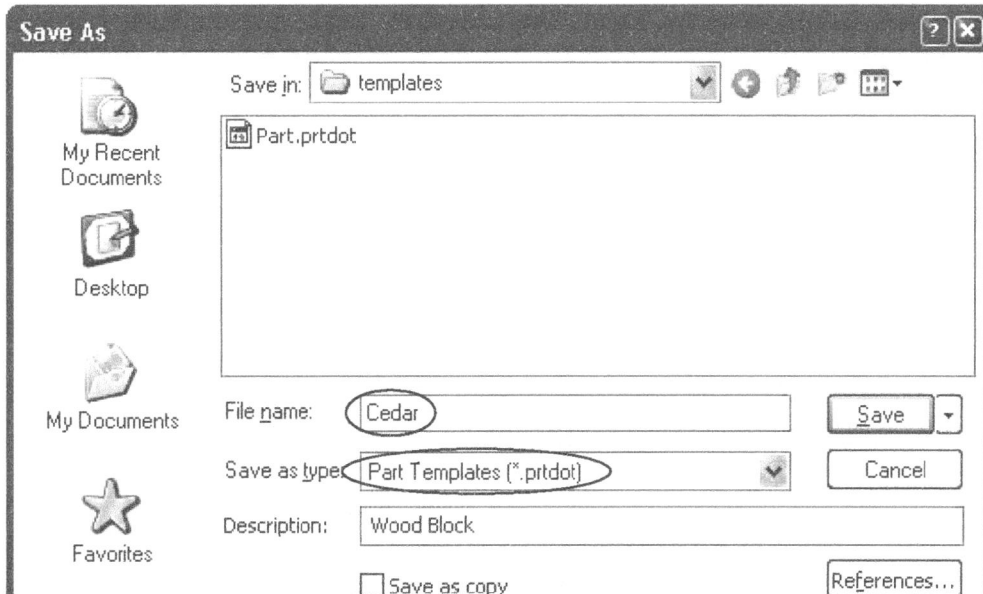

Pull down the "File" menu and pick **Close**. Once a template is saved, you can use it to start a new part. The new part will contain the specific settings that you saved to that template. SolidWorks lets you create as many template files for each document type as you want. Using templates can give a big head start when creating new parts or assemblies.

Change the Default Template Files

When SolidWorks creates a new file and doesn't prompt for a template file, SolidWorks uses the settings stored in the default template files. The default templates are used for automatically created parts, assemblies, and drawings. For example, when you import a file from another application or create a derived part, the default template is used for the new document.

To change which templates SolidWorks uses, pull down the "Tools" menu and pick **Options**.

On the **System Options** tab, click on **Default Templates**.

To change the default, click on the [...] button to the right of the document type you want and browse to the desired template file, and then click **OK**. Note that below this area is an option which gives you a choice of using the default document templates or being prompted each time.

System Options - Default Templates

System Options | Document Properties

- General
- Drawings
  - Display Style
  - Area Hatch/Fill
- Colors
- Sketch
  - Relations/Snaps
- Display/Selection
- Performance
- Assemblies
- External References
- Default Templates
- File Locations
- FeatureManager
- Spin Box Increments

These templates will be used for operations (such as File Import and Mirror Part) where SolidWorks does not prompt for a template.

Parts

C:\Program Files\SolidWorks\data\templates\Cedar.prtdot

Assemblies

Drawings

⦿ Always use these default document templates
◯ Prompt user to select document template

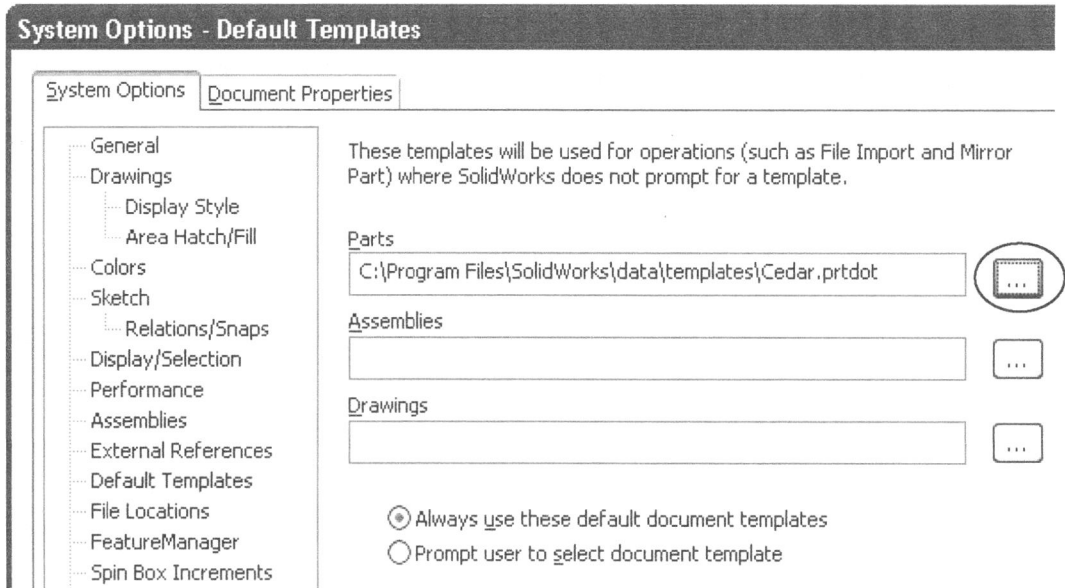

Next, on the **System Options** tab, click on **File Locations** and check that the default file locations are as shown below. Since you just saved a template into the templates folder, it should appear here in the list of folders. If you don't have the templates folder in the list, click the **Add** button and browse to select the folder. These folder locations are where SolidWorks looks to find and display template files. You can create and name your own templates folder and save your templates wherever you want. Just make sure that you add the location here so that SolidWorks can find your template files. When you're done, click **OK**.

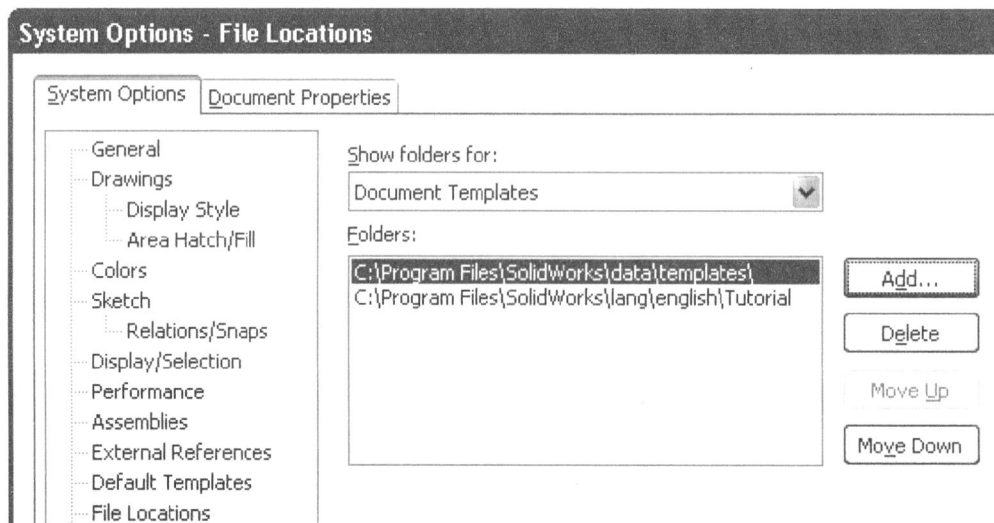

System Options - File Locations

System Options | Document Properties

- General
- Drawings
  - Display Style
  - Area Hatch/Fill
- Colors
- Sketch
  - Relations/Snaps
- Display/Selection
- Performance
- Assemblies
- External References
- Default Templates
- File Locations

Show folders for:

Document Templates

Folders:

C:\Program Files\SolidWorks\data\templates\
C:\Program Files\SolidWorks\lang\english\Tutorial

Add...
Delete
Move Up
Move Down

Use a Custom Template

You can use a custom template when creating a new document file. This allows you to create a new file by using a previously saved template.

Press **Ctrl+N**, or pull down the "File" menu and pick **New**.

In the **New SolidWorks Document** dialog box, you will see the **File Locations** added in the previous section displayed as tabs in the dialog box. This will only appear in the advanced **New SolidWorks Document** dialog box. If you are looking at the novice dialog box, you will see an **Advanced** button in the lower left of the dialog box. Click the **Advanced** button. [Advanced]

On the **Templates** tab, you will see your custom template. Pick **Cedar**, and click **OK**. When you pick a custom template, a new document will be created using the properties of the template.

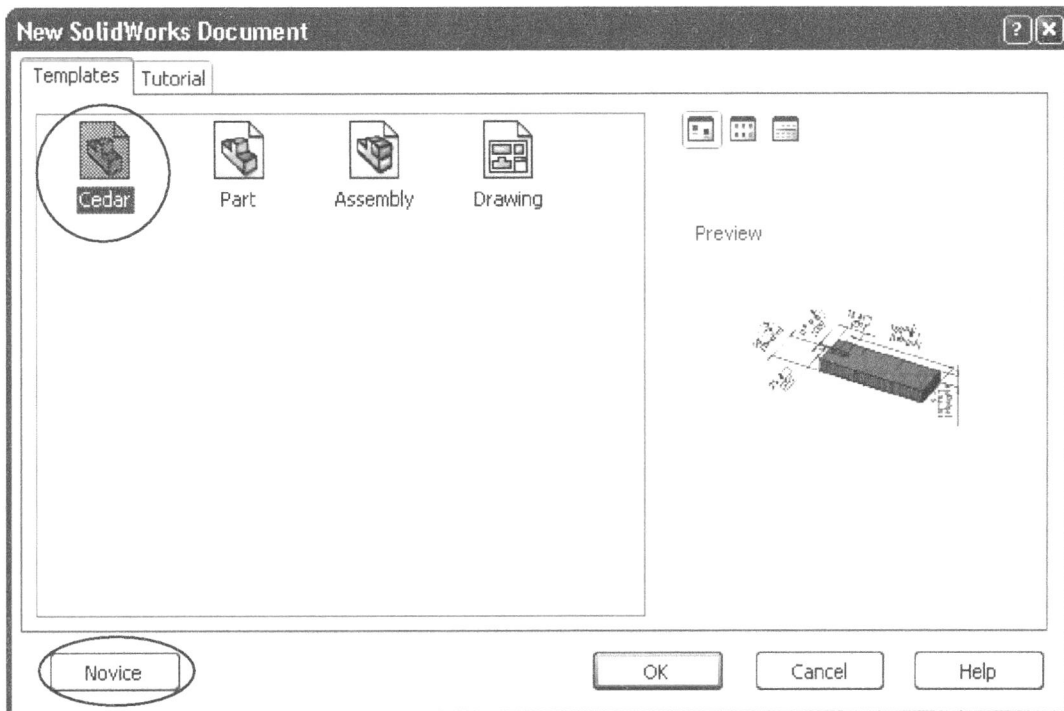

Double click on the **150 [5 7/8"] (Length)** dimension and change it to '**75**' and press the **Enter** key.

Then, press **Ctrl+Shift+B** to rebuild the model and change to the Isometric view. After it rebuilds the model, it displays a message asking if you want to **Change to Isometric View?** Click **No**. Remember, this is what you created the macro to do in Chapter 6.

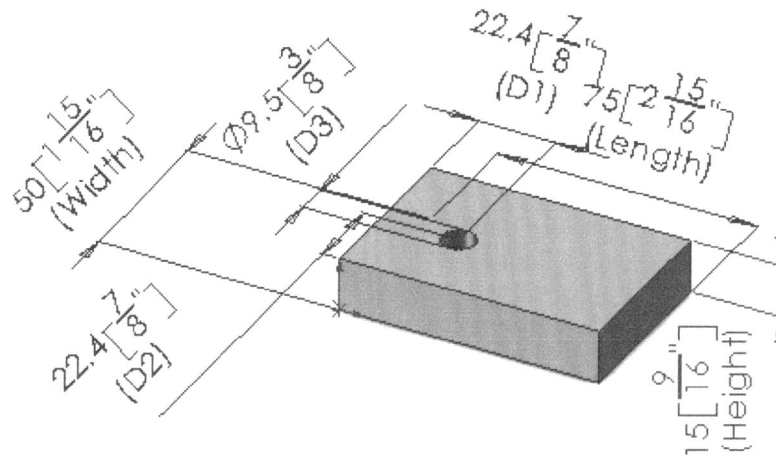

If this were a real part you were creating, you would continue to make changes and add feature to the model. I just wanted to show you how to get started.

Pull down the "Window" menu and pick **Close All**.

Pick **Yes** to save changes. In the **Save as** dialog box, browse to the default folder where you save SolidWorks part files and type '**8463-38**' (This will be used for your part number in the next chapter) for the **File name** and click **Save**.

Chapter 11

Creating your own drawing template lets you standardize the drawing paper format and title block information for all your drawings. When you start a new drawing, SolidWorks loads the default setup values of the selected template. In most companies, the drawing standards are the same for every drawing produced. The border and title block have a standard that must be maintained. You will save considerable time if you set this up once and save all these values rather than having to change them for every new drawing.

That's where custom sheet formats and drawing templates come in. You can very easily create multiple drawing templates that contain all the initial setup information configured according to your specifications. The sheet formats can contain a border, title block, specific layers, and even a set of standard notes that are common to all drawings. Learn how to automatically fill in information from the part or assembly document into the drawing by linking notes to properties.

Create a New Drawing Document

Click the **New** button in the "Standard" toolbar, or pull down the "File" menu and pick **New**.

The **New SolidWorks Document** dialog box appears.

Click the **Drawing** icon and then click **OK**.

Drawing

In the **Sheet Format/Size** dialog box, make sure that the **Standard sheet size** radio button is selected and pick **A - Landscape** from the menu. Right below the menu is the name of the template, **a - landscape.slddrt**.

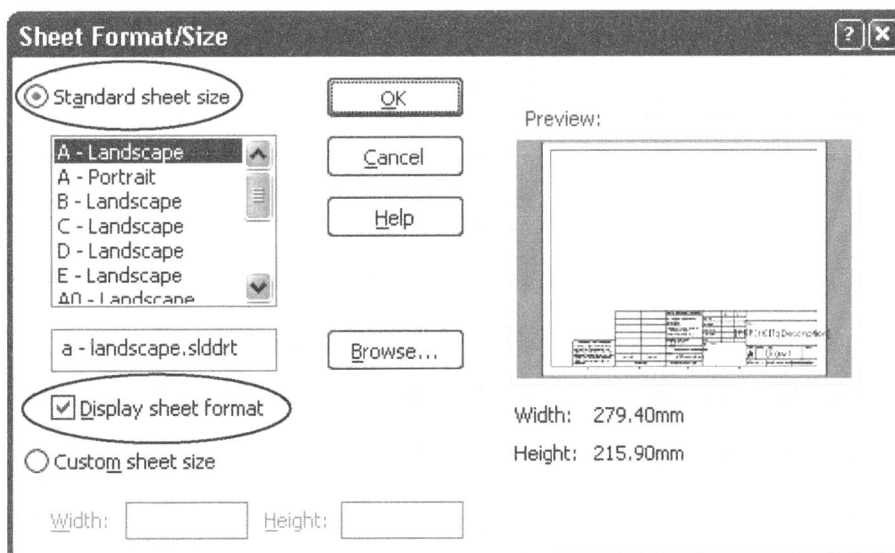

Make sure that **Display sheet format** is checked and click **OK**. This will include the border and the title block in your drawing. Otherwise you would be looking at a blank piece of paper.

In the **Model View** PropertyManager, click the **Cancel** button.

Prepare the Drawing Template Format

SolidWorks allows you to customize the drawing sheet format by changing the properties. By default, the SolidWorks template has some built-in fields that will fill in automatically for you. Also, other notes have been located and formatted for your use that are linked to the drawing properties.

To customize the drawing sheet format, right click anywhere on the drawing sheet and pick **Edit Sheet Format** from the menu. The lines on the title block and border turn blue, indicating that the drawing sheet format is now active, or selectable. Note that you can change the size and location of the blue sketch lines, as well as create new sketch lines. This may be necessary to move the border lines within the margins of your plotter or to create you own title block.

Click the **Zoom to Area** button in the "View" toolbar, or pull down the "View" menu and pick **Modify – Zoom to Area**, and zoom into the lower right corner of the title block as shown.

Press the **Escape** key. If you hover the cursor over the middle of the box above **Title**, you will notice the note text **$PRP:"COMPANYNAME"**.

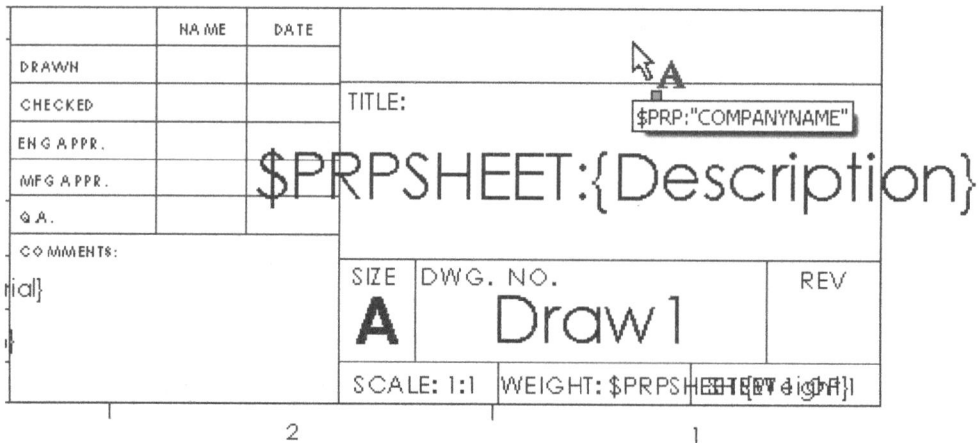

Pull down the "File" menu and pick **Properties**.

In the **Summary Information** dialog box, on the **Custom** tab, click in the box below **Property Name** and pull down the "Property Name" menu and pick **CompanyName**. The **Type** should fill in to say **Text**. If not, pull down the list and pick **Text**.

Click in the **Value / Text Expression** box and type your company name just the way you want it to appear in the title block.

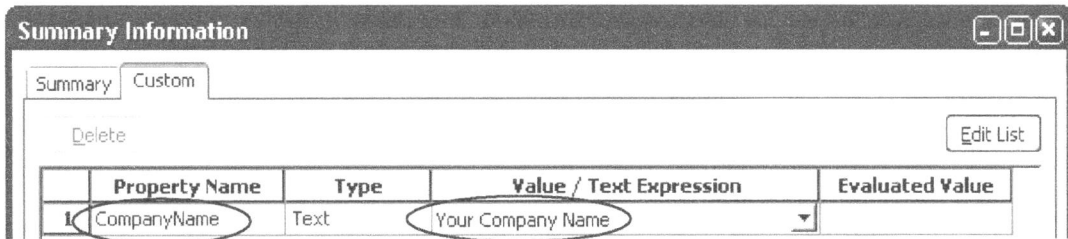

Click **OK**. Note that the **CompanyName** text is automatically placed on your drawing.

Double click on your company name. The **Formatting** toolbar appears along with the **Note** PropertyManager. Pick a font and point size for your company name.

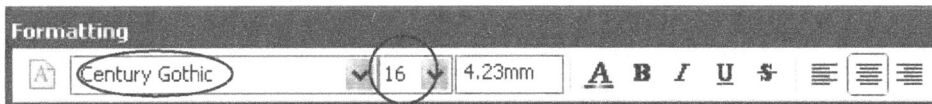

Click the green check mark button at the top of the **Note** PropertyManager when you are finished.

Link to Property

SolidWorks lets you link the value of a document property, a custom property, or a configuration specific property to the text of a note. If you do this, you don't have to worry if you change the property because the note text will automatically change for you. To see what I mean, click on the note text in the **Drawn/Name** box as shown (A little green box will appear where you click and the cursor flyout will say **$PRP:"DrawnBy"**).

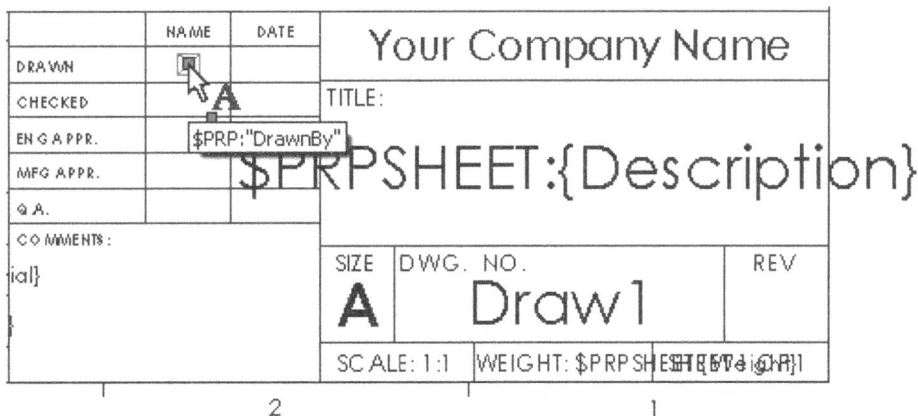

In the **Note** PropertyManager, under **Text Format**, click the **Link to Property** button.

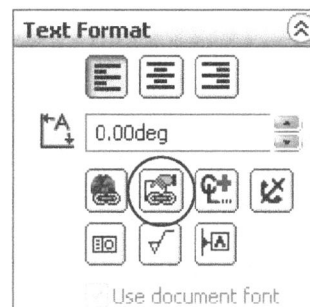

In the **Link to Property** dialog box, click the **Model in view specified in sheet properties** radio button.

Click in the text box and type '**DrawnBy**', exactly as it appears here, and click **OK**.

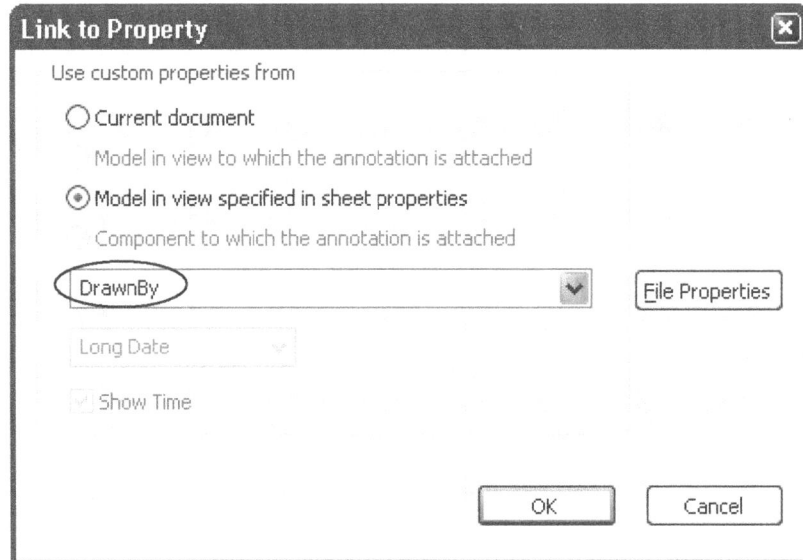

Click the green check mark button at the top of the **Note** PropertyManager.

Now click on the note text in the **Drawn/Date** box as shown (A little green box will appear where you click and the cursor flyout will say **$PRP:"DrawnDate"**).

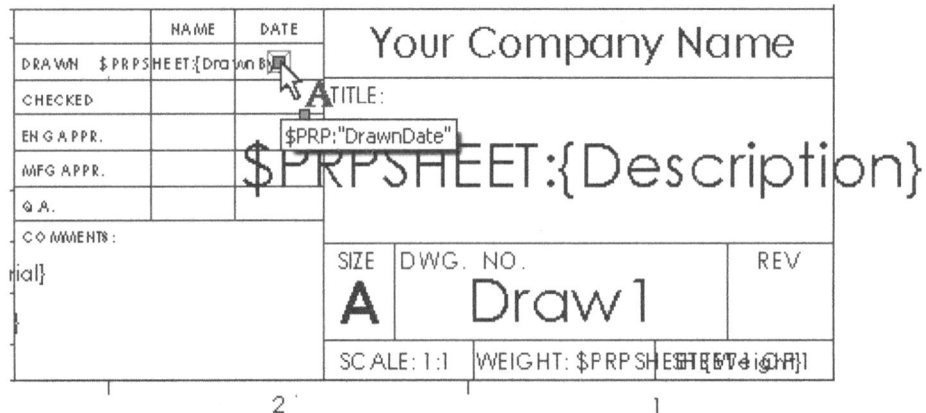

In the **Note** PropertyManager, under **Text Format**, click the **Link to Property** button.

In the **Link to Property** dialog box, click the **Model in view specified in sheet properties** radio button.

Pull down the menu and pick **SW-Short Date**, and click **OK**.

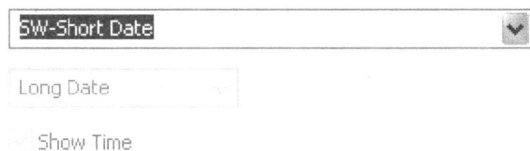

Click the green check mark button at the top of the **Note** PropertyManager. The **Link to Property** button is available for notes that you create as well.

Next, double click on the shop tolerance statement in the title block, as shown.

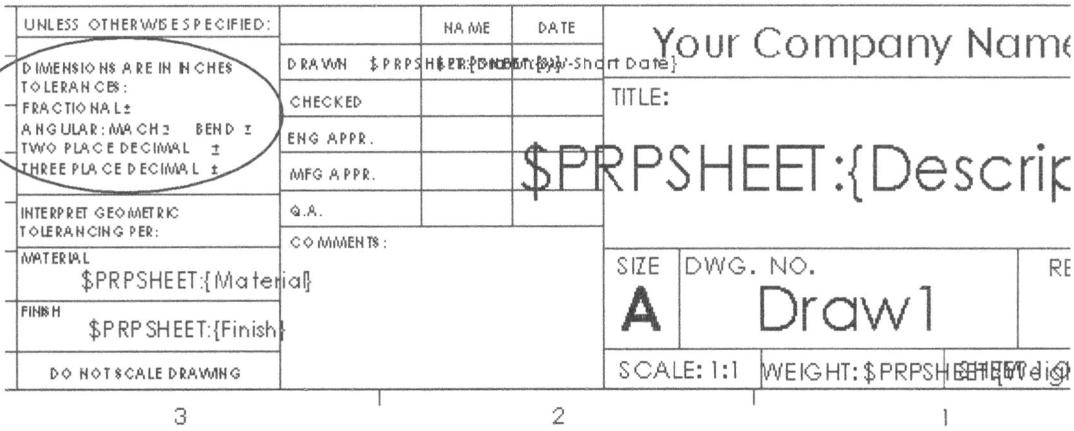

Edit the text by replacing **INCHES** with '**mm**', deleting **FRACTIONAL**, adding an angular tolerance of '**0.01**', and deleting **BEND**. Change **TWO** to '**ONE**' and add '**0.05**'. Then, change **THREE** to '**TWO**' and add '**0.005**'.

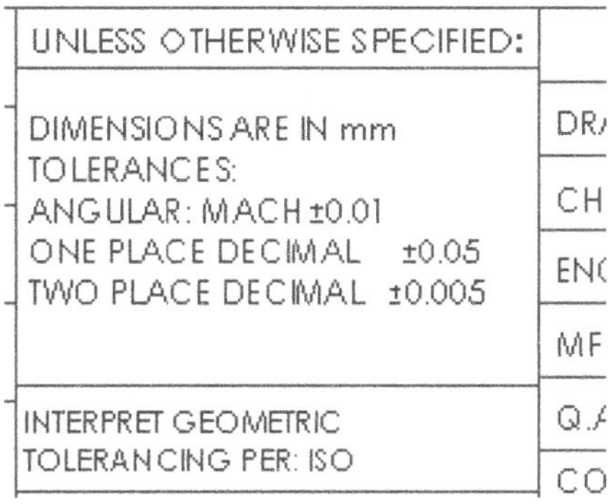

Click the green check mark button at the top of the **Note** PropertyManager. You can also just click anywhere outside the text box.

Double click on the **INTERPRET GEOMETRIC TOLERANCING PER:** text and add '**ISO**' as shown.

Then, click the green check mark button at the top of the **Note** PropertyManager.

Press the **F** key to **Zoom to Fit**, or pull down the "View" menu and pick **Modify – Zoom to Fit**.

Right click in the graphics area and pick **Edit Sheet** from the menu. The lines on the title block turn gray, indicating that the drawing sheet is now active.

Next, save the updated drawing sheet format by pulling down the "File" menu and picking **Save Sheet Format**.

In the **Save Sheet Format** dialog box, browse to the **<SolidWorks Install Directory>/data** directory change and the **File name** to 'a - **<Your Company Name> landscape**'. Since the dialog box to open a new drawing document has limited space for displaying the sheet format name, you may want to abbreviate your company name. For example, '**a - YCN landscape**'.

Click **OK**. When you choose this format for your drawings, you will not need to make your changes again.

Set the Detailing Options

There are many options under detailing that you can set up ahead of time and save in your template. SolidWorks allows you to choose the default dimension standard, dimension font, the style of dimensions, arrows, and other detailing options. For now, use the settings described in this chapter. Later, set the detailing options to the way that you or your company wants them. You may want to set up multiple drawing templates for different units or paper sizes.

In the FeatureManager design tree, right click on **Draw1** and pick **Document Properties**. This option will appear if you click anywhere in the FeatureManager design tree with nothing selected.

On the **Document Properties** tab, in the **Dimensioning standard** section, make sure that **ISO** is selected. Then, pull down the **Trailing zeroes** menu and pick **Remove**.

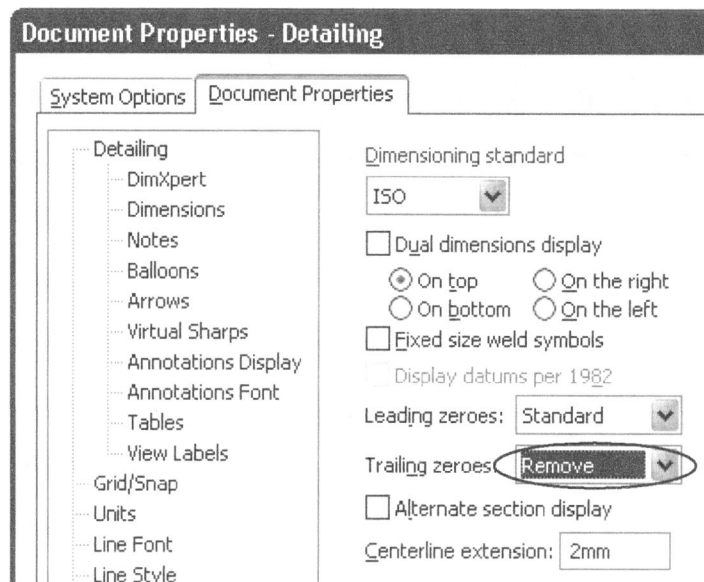

Document Properties - Detailing

System Options | Document Properties

- Detailing
  - DimXpert
  - Dimensions
  - Notes
  - Balloons
  - Arrows
  - Virtual Sharps
  - Annotations Display
  - Annotations Font
  - Tables
  - View Labels
- Grid/Snap
- Units
- Line Font
- Line Style

Dimensioning standard

ISO

☐ Dual dimensions display
  ◉ On top   ◯ On the right
  ◯ On bottom   ◯ On the left
☐ Fixed size weld symbols
   Display datums per 1982

Leading zeroes: Standard

Trailing zeroes: Remove

☐ Alternate section display

Centerline extension: 2mm

On the **Document Properties** tab, under **Detailing**, click **Annotations Font**.

Under **Annotation type**, click **Dimension**.

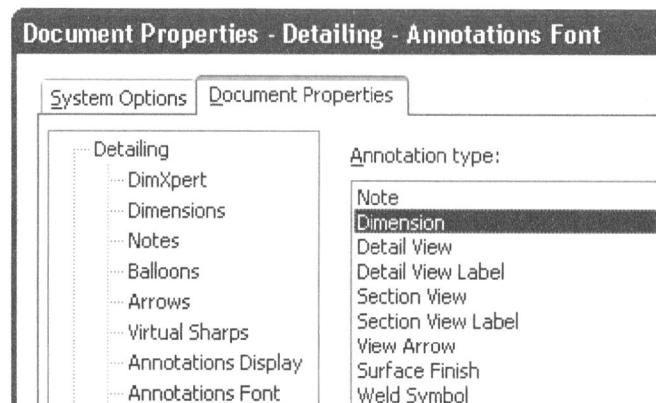

Document Properties - Detailing - Annotations Font

System Options | Document Properties

- Detailing
  - DimXpert
  - Dimensions
  - Notes
  - Balloons
  - Arrows
  - Virtual Sharps
  - Annotations Display
  - Annotations Font

Annotation type:

Note
Dimension
Detail View
Detail View Label
Section View
Section View Label
View Arrow
Surface Finish
Weld Symbol

In the **Choose Font** dialog box, in the **Height** section, click the **Points** radio button and select **16**.

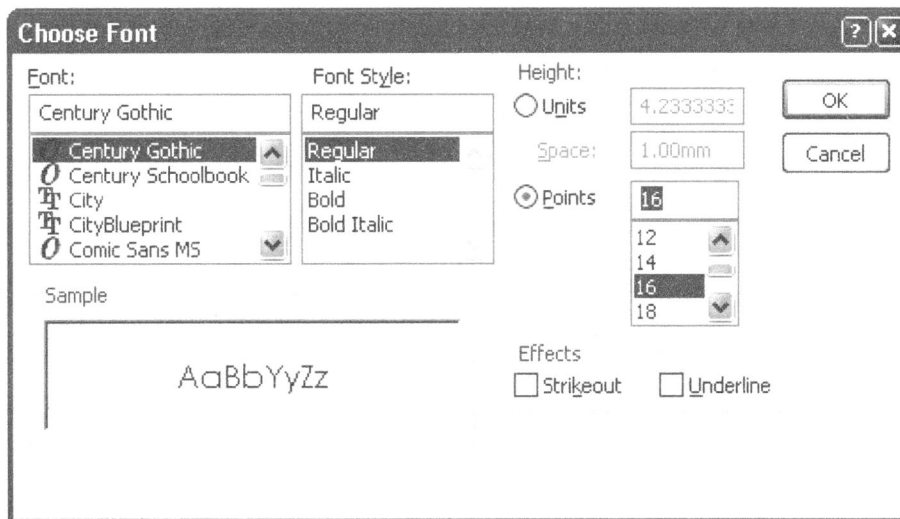

In the **Choose Font** dialog box, click **OK**.

On the **Document Properties** tab, click **Units**.

Set the **Unit** system to **MMGS (millimeter, gram, second)** and the **Dual units** to **Inches**.

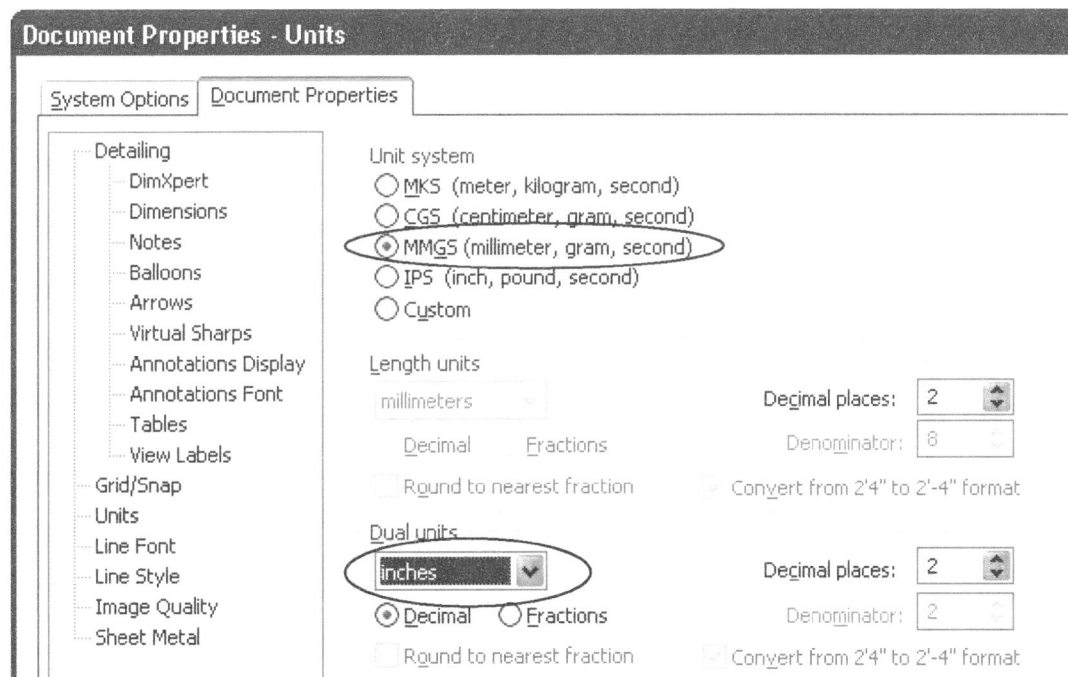

Then, in the **Document Properties** dialog box, click **OK** to close the dialog box.

Set Up Layers

Layers can be used in a drawing document to easily assign a line color, line thickness, and line style for new entities created on each layer. Setting up these layers ahead of time is a simple procedure.

Right click on a toolbar, not the CommandManager, and pick **Layer** to turn on the "Layer" toolbar.

Click the **Layer Properties** button in the "Layer" toolbar.

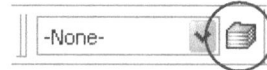

In the **Layers** dialog box, click the **New** button, and enter '**Dimensions**' for the **Name** of a new layer.

Double click in the **Description** column and type '**Reference**'.

To specify the color, click the **Color** box, pick **Red**, and click **OK**.

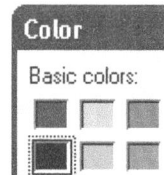

In the **Layers** dialog box, click **OK**.

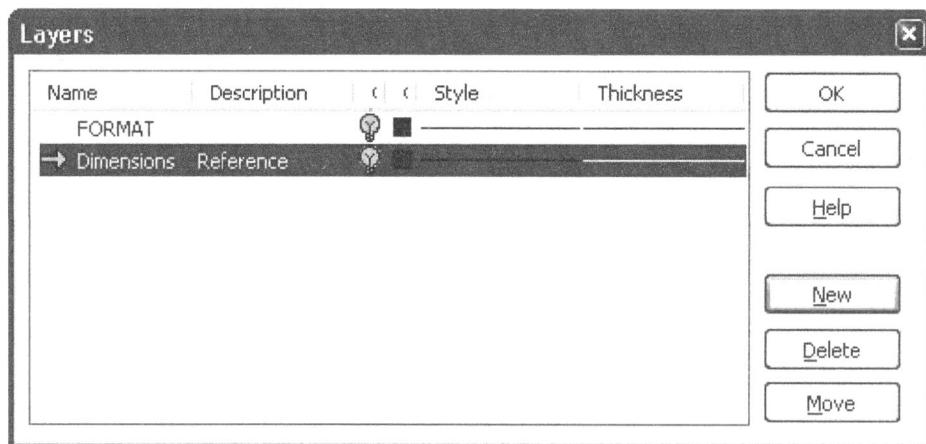

The **Dimensions** layer is the active layer. Any new entities that are created when this layer is active will be the color red. Change the active layer to **-None-** by pulling down the menu and picking -None- in the "Layer" toolbar.

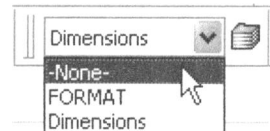

Save the Part

Now you're ready to save this drawing as a drawing template so that it can used over and over to create new drawings.

Press **Ctrl+S**, or pull down the "File" menu and pick **Save**.

In the **Save As** dialog box, in the **File name** box, type '**Custom Drawing**' for the name of the drawing template.

Under **Save as type**, pick **Drawing Templates (*.drwdot)**. The folder location should change to the **<SolidWorks Install Directory>/data/templates** folder automatically.

Click **Save**.

Pull down the "File" menu and pick **Close**.

Try it Out

Press **Ctrl+N** or click the **New** button in the "Standard" toolbar.

The **New SolidWorks Document** dialog box appears.

Double click on the **Custom Drawing** icon on the **Templates** tab.

In the **Model View** PropertyManager, click on the **Browse** button.

Browse to the SolidWorks Install Directory and double click on **8463-38.SLDPRT**.

Click in the graphics area to place a view.

Click the green check mark button at the top of the **Projected View** PropertyManager.

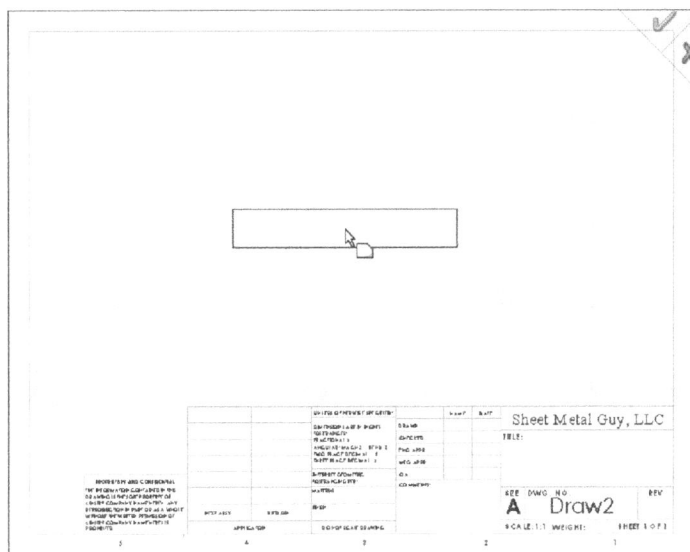

Click the **Zoom to Area** button in the "View" toolbar, or pull down the "View" menu and pick **Modify – Zoom to Area**, and zoom in to the lower right corner of the title block. The custom properties that were assigned to the part now automatically populate the title block.

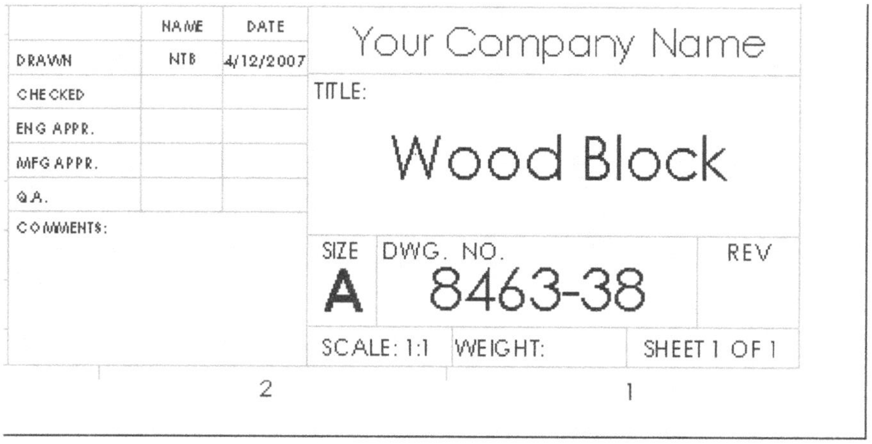

	NAME	DATE	Your Company Name			
DRAWN	NTB	4/12/2007				
CHECKED			TITLE:			
ENG APPR.						
MFG APPR.			**Wood Block**			
Q.A.						
COMMENTS:						
			SIZE	DWG. NO.		REV
			A	**8463-38**		
			SCALE: 1:1	WEIGHT:	SHEET 1 OF 1	
	2			1		

Press the **Escape** key.

Right click in the graphics area and pick **Properties**.

In the **Sheet Properties** dialog box, note that **A – YCN landscape** is selected.

Then, click **OK**.

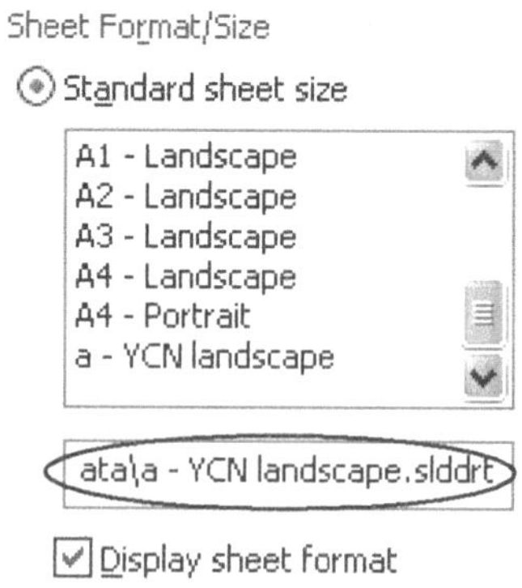

Pull down the "File" menu and pick **Close**.

Click **No** to not save the file. This is a sample drawing document. There is no reason to save it.

Now, create your own templates. This chapter used an A-sized sheet. You'll want to create similar templates with other units or sheet sizes that you use. The templates are also available when you use the **Make Drawing from Part/Assembly** command.

Sheet Format/Size

⦿ Standard sheet size

A1 - Landscape
A2 - Landscape
A3 - Landscape
A4 - Landscape
A4 - Portrait
a - YCN landscape

ata\a - YCN landscape.slddrt

☑ Display sheet format

Chapter 12

Collaboration with Others

Collaboration is when more than one user of SolidWorks is working on the same project simultaneously. In an ideal world, changes would automatically appear on everyone's screen in a real-time mode. But collaboration doesn't work that way. The best method for working together is to have all of the files saved in one common location, a network drive, and each user takes sole write access of the file they are working on to reduce the chance of someone else changing the same file or other related documents.

If you are working on a shared network drive without a product data management (PDM) system, you will benefit from the collaboration options offered by SolidWorks. The SolidWorks program offers some basic features to help you collaborate with others. If you need more advanced features, you may need to consider investing in a product data management (PDM) system.

Those users actively editing parts need to be reminded to save their work on a regular basis and those users with read-only reference files open need to be reminded to reload the modified files so that they are viewing the most current files.

To work well with one another in a group, users must set several settings to allow the sharing of files and updating of files. You also need some form of communication between group members to accommodate discussion as issues arise. You need to pay attention to the read / write access of a file before you make changes. If you do not have write access to a file, you can still make changes to the read-only file. The changes, though, can only be saved to a new file name to make sure that changes are not lost or override the other user's changes.

Set Collaboration Options

To get started, you need to specify a few options in SolidWorks for a multi-user environment.

Pull down the "Tools" menu and pick **Options**.

On the **System Options** tab, pick **Collaboration**, and then check **Enable multi-user environment** to enable the other options.

Check **Add shortcut menu items for multi-user environment** to add special collaboration menu items. With this option checked, **Make Read-Only** and **Get Write Access** are available on the "File" pull down menu for assembly and part documents, and when you right click assembly components in the FeatureManager design tree or in the graphics area. When more than one person has a document file open, only one person can have write access (allowing you to save the current file) for that document.

Chapter 12: Collaboration with Others   139

Click **Check if files opened read-only have been modified by other users**. This will check files you have open as read-only every 20 minutes, the interval specified in **Check files every xx minutes**, to see if the files have been modified. SolidWorks will tell you if your file is out of date if another user has saved a file that you have open in SolidWorks. SolidWorks will also check to see if another user has relinquished write access to a file that you have open, allowing you to take write access.

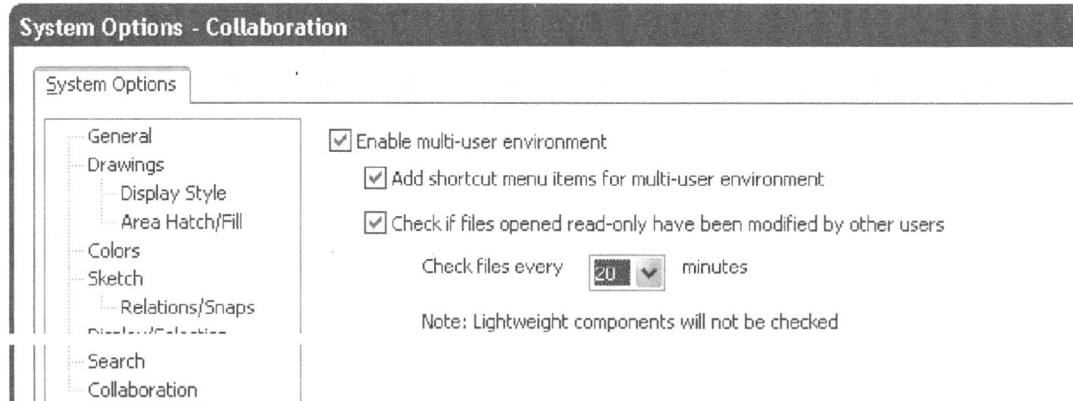

System Options - Collaboration

System Options

- General
- Drawings
  - Display Style
  - Area Hatch/Fill
- Colors
- Sketch
  - Relations/Snaps
- Search
- Collaboration

☑ Enable multi-user environment

  ☑ Add shortcut menu items for multi-user environment

  ☑ Check if files opened read-only have been modified by other users

    Check files every   **20** ▾   minutes

    Note: Lightweight components will not be checked

On the **System Options** tab, pick **External References**.

Check **Open referenced documents with read-only access** and **Don't prompt to save read-only referenced documents (discard changes)** to insure that files that you did not want to change are loaded as read-only and not saved accidentally.

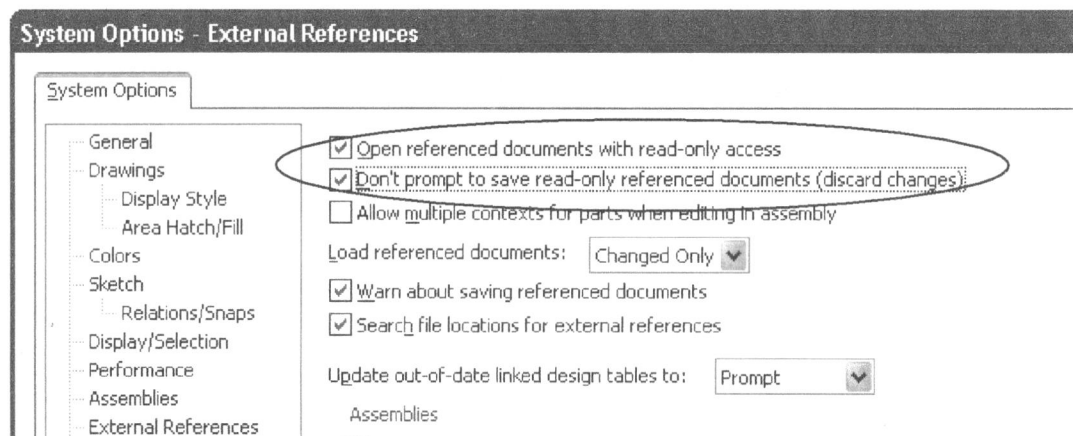

System Options - External References

System Options

- General
- Drawings
  - Display Style
  - Area Hatch/Fill
- Colors
- Sketch
  - Relations/Snaps
- Display/Selection
- Performance
- Assemblies
- External References

☑ Open referenced documents with read-only access
☑ Don't prompt to save read-only referenced documents (discard changes)
☐ Allow multiple contexts for parts when editing in assembly

Load referenced documents: Changed Only ▾

☑ Warn about saving referenced documents
☑ Search file locations for external references

Update out-of-date linked design tables to: Prompt ▾

  Assemblies

In the **System Options** dialog box, click **OK**.

Network File Locations

Multiple users should store their SolidWorks files in one common location, preferably on a network drive, rather than in many separate locations. This way, there is no confusion and there are no duplicate files scattered over multiple computers. You can easily set up a network location for the SolidWorks files to be stored. This is usually a specific folder located on a network drive that everyone has access to.

To do this, open Windows Explorer. You may need to get your system administrator involved to ensure that everyone has the proper level of permission to access this folder.

Browse to a network drive and create a new folder named 'SolidWorks Files'.

Using Windows Explorer, browse to the default **<SolidWorks Install Directory>\data\templates** folder and move **cedar.prtdot** and **Custom Drawing.drwdot** to your new **SolidWorks Files** folder on the network drive.

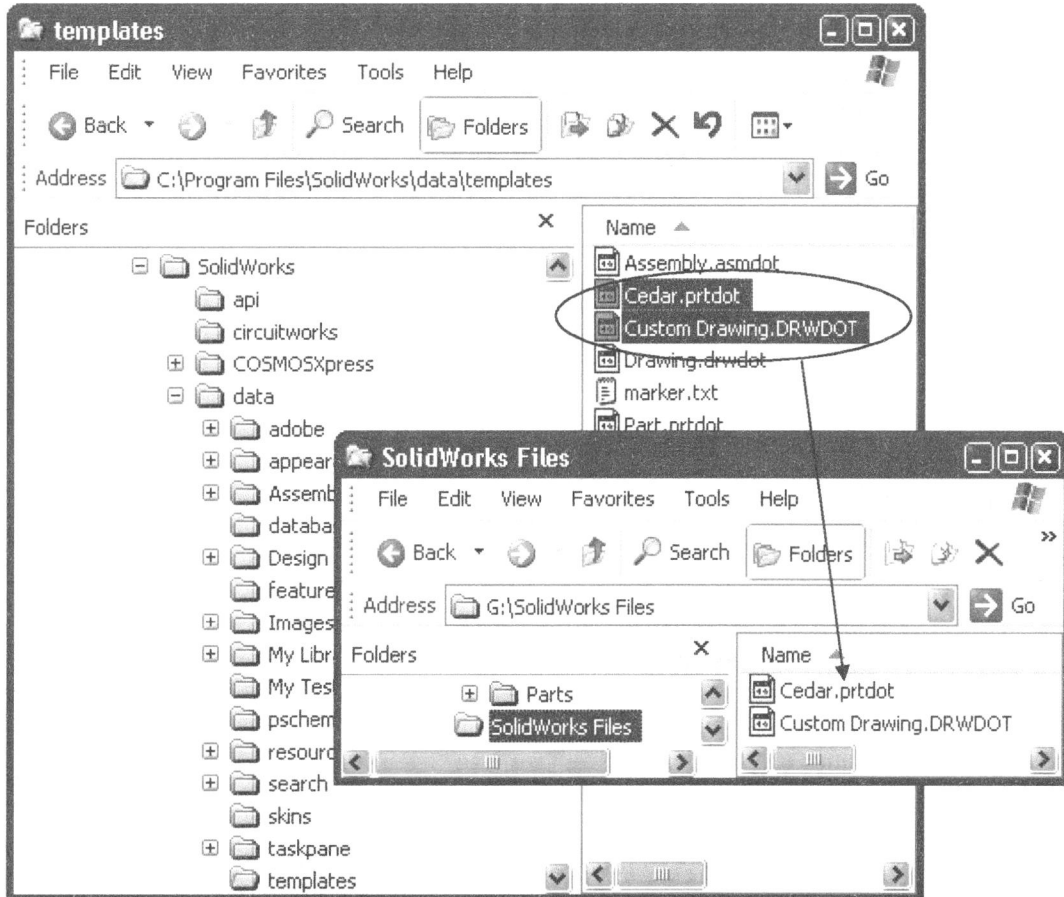

In SolidWorks, pull down the "Tools" menu and pick **Options**.

On the **System Options** tab, pick **File Locations**.

Under **Show folders for**, with **Document Templates** selected, click the **Add** button.

In the **Browse For Folder** dialog box, browse to the **SolidWorks Files** folder, and then, click **OK**.

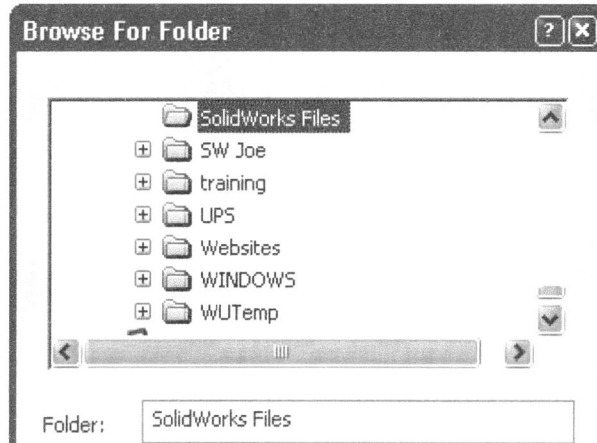

The **SolidWorks Files** folder now appears in the list of folders for **Document Templates**.

Click on the **SolidWorks Files** folder to highlight it and click the **Move Up** button to move the **SolidWorks Files** folder up on the list. This will determine the order of the tabs on the **New SolidWorks Document** dialog box. If it exists, the templates folder is always the first tab.

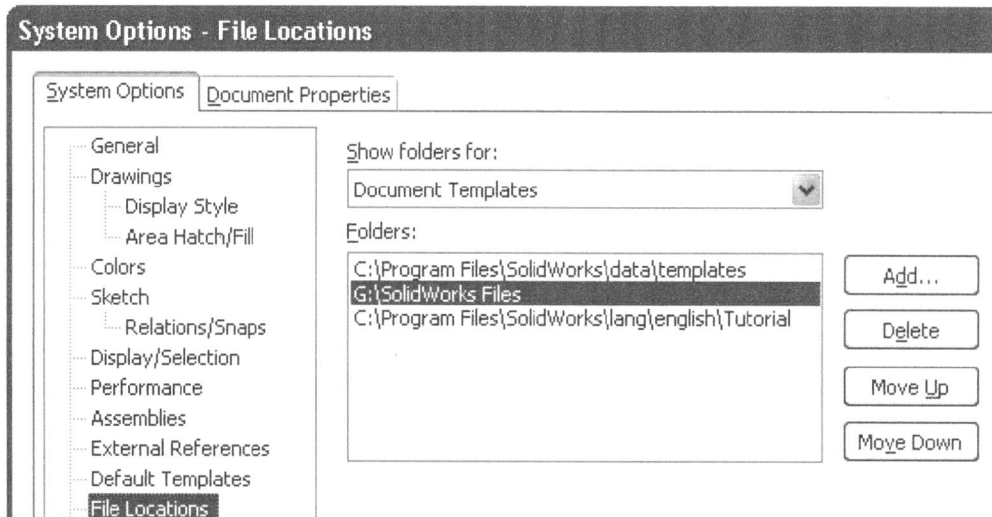

In the **System Options** dialog box, click **OK**.

Press **Ctrl+N**, or click the **New** button in the "Standard" toolbar.

Click on the **SolidWorks Files** tab. Your cedar template and custom drawing template files should still appear, but they are now located on the network drive instead of your local drive, allowing access by all users.

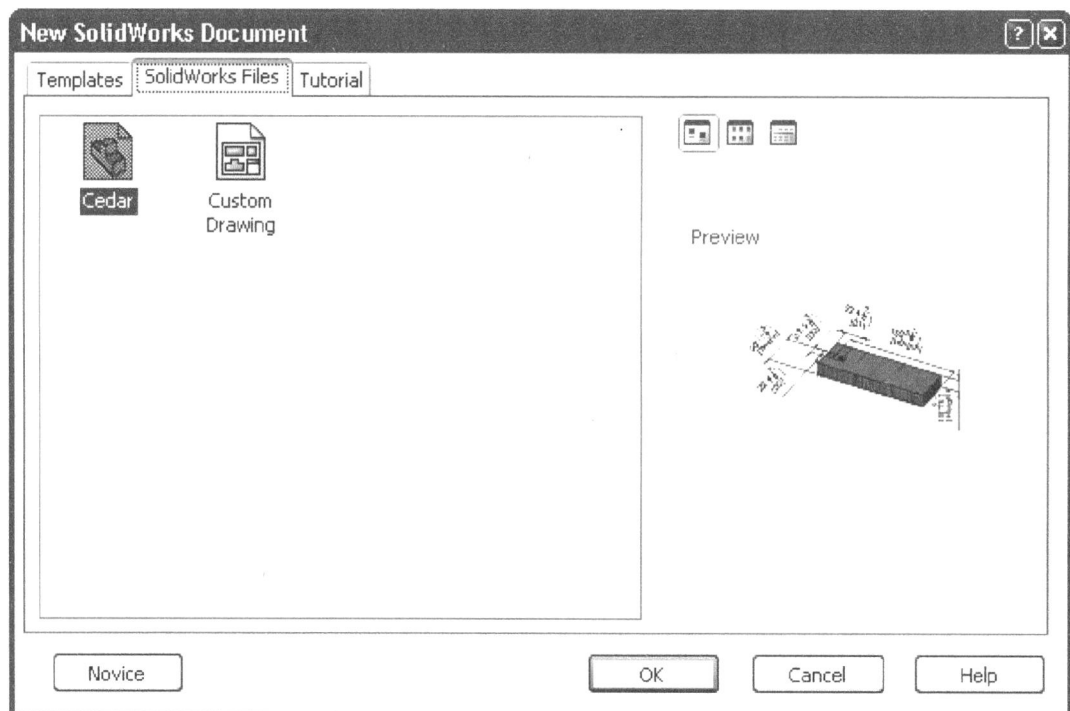

Double click on **Cedar** to open the file.

Change the dimensions as shown and press **Ctrl+B** to rebuild the model.

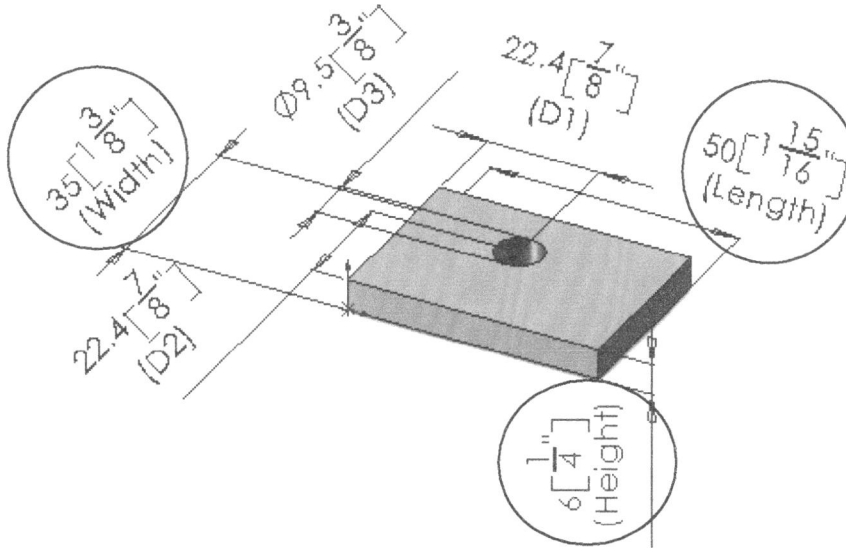

Press **Ctrl+S**, or pull down the "File" menu and pick **Save**.

In the **Save As** dialog box, enter '**Block**'.

Browse to the **SolidWorks Files** folder on the network drive and click **Save** to store the file in a location that can be easily accessed by others.

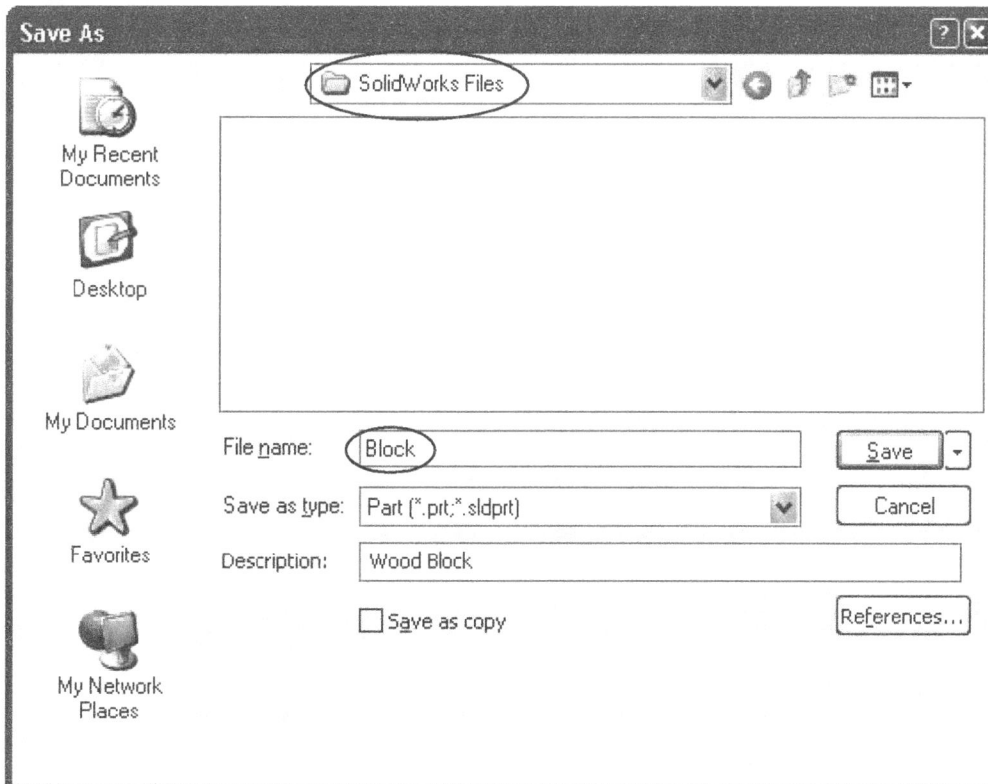

Pull down the "File" menu and pick **Close**.

Press **Ctrl+O**, or pull down the "File" menu and pick **Open**.

In the **Open** dialog box, double click on **Block** in the **SolidWorks Files** folder.

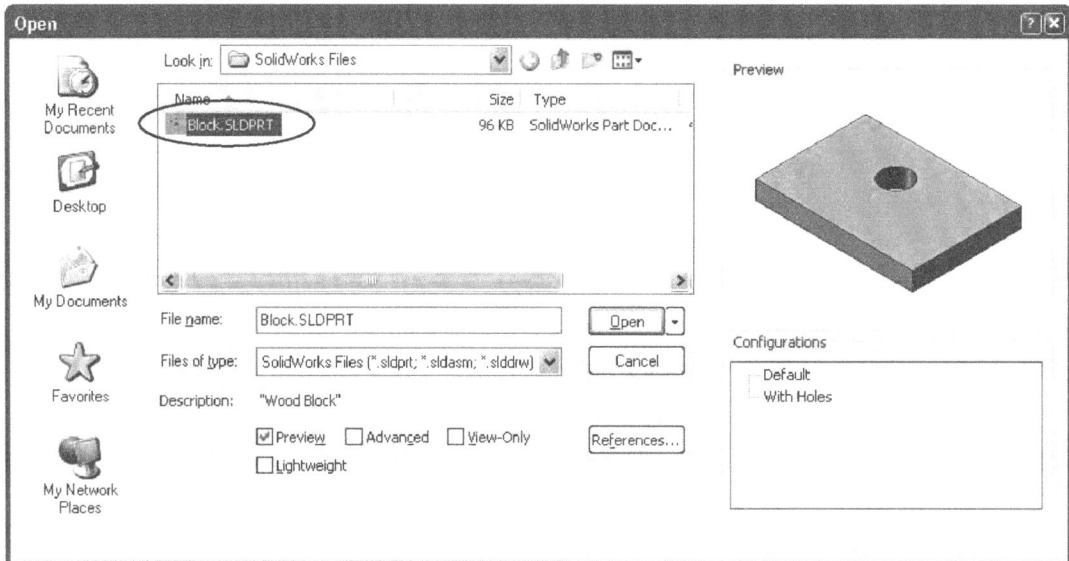

Get Write Access or Make Read-Only

Pull down the "File" menu and pick **Make Read Only**. The title bar changes to show that the file is now read-only.

Pull down the "File" menu and pick **Get Write Access**. The title bar changes back to show that the file is no longer read-only.

Note that assembly components can either be read-only or write-access enabled. This setting can be changed at any time by right clicking on the component in the FeatureManager design tree and picking either **Get Write Access** or **Make Read-Only**. If a file is already open by someone else, SolidWorks will prompt you to open a copy, as shown below. If you pick **Yes**, a read-only copy will be opened. If you pick **No**, the file will not be opened.

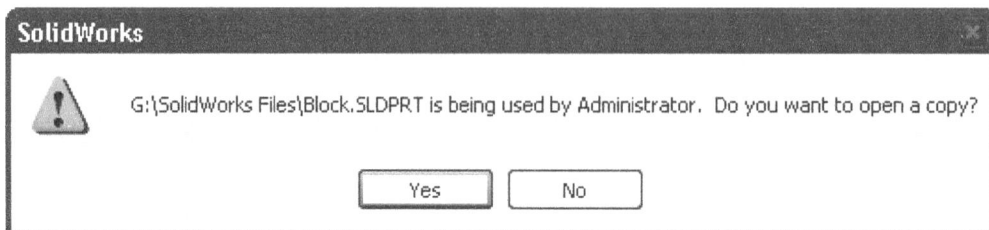

Manually Check File Status

SolidWorks also has an option to check to see if others have updated one of the documents in your current assembly. You can manually check the file status at any time by adding the **Check read-only files** button to your "Explode Sketch" toolbar.

To do this, click the **Explode Sketch** button in the control area of the CommandManager.

Then, press **Ctrl+Alt+C**, or pull down the "Tools" menu and pick **Customize**.

Click on the **Commands** tab, and then, under **Categories**, pick **Standard**. Under **Buttons**, locate the **Check read-only files** button and drag the button into your "Explode Sketch" toolbar in the CommandManager, as shown.

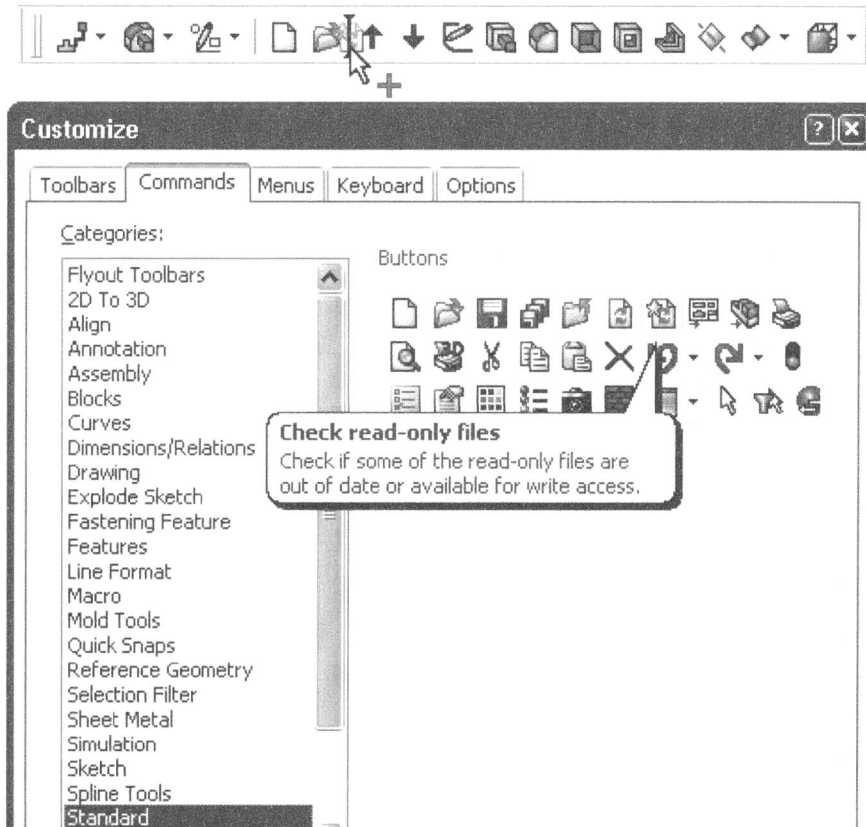

In the **Customize** dialog box, click **OK**.

Click the **Check read-only files** button in the "Explode Sketch" toolbar to check to see if any of the read-only documents were updated.

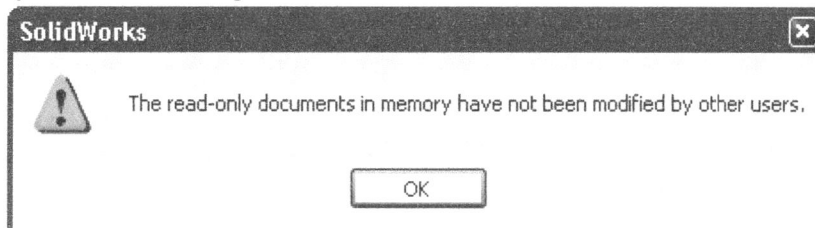

If SolidWorks detects a change, a tooltip in the lower right of the graphics area points to a button on the status bar, as shown below. SolidWorks will check if files opened read-only have been modified by other users at the interval you specified at the beginning of the chapter (20 minutes).

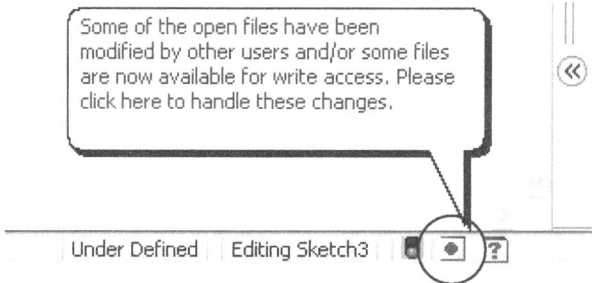

Some of the open files have been
modified by other users and/or some files
are now available for write access. Please
click here to handle these changes.

Under Defined   Editing Sketch3

When you click the **Reload** button below the tooltip, the **Reload** dialog box appears. You may also access the **Reload** dialog by pulling down the "File" menu and picking **Reload**. The **Reload** dialog box allows you to select components to reload modified items into the current session or to reload the original version of a document without exiting SolidWorks. You can also change write access.

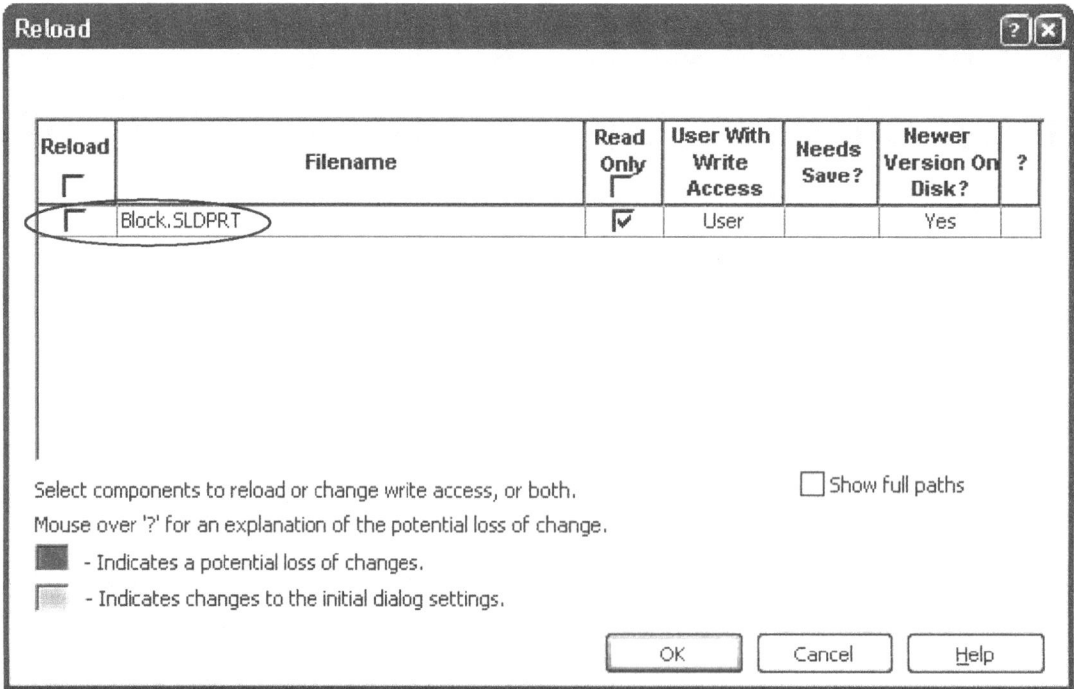

Reload

Reload	Filename	Read Only	User With Write Access	Needs Save?	Newer Version On Disk?	?
	Block.SLDPRT	☑	User		Yes	

Select components to reload or change write access, or both.

Mouse over '?' for an explanation of the potential loss of change.

- Indicates a potential loss of changes.

- Indicates changes to the initial dialog settings.

☐ Show full paths

OK   Cancel   Help

Pull down the "File" menu and pick **Close**.

This should get you started in the right direction. You will have to come up with the best method of collaboration that will work within your organization. Consider your company procedures and develop a standard to implement throughout all the SolidWorks users in your organization. There is no right or wrong way. Just make sure that you are consistent and that everyone is on the same page to make working together problem free.

The next chapter will help you see ways to set up everyone's work station the same.

Chapter 13

The Copy Settings Wizard can be used to save and backup your settings and share those settings with others without having to edit the Windows registry. This allows for SolidWorks to be configured on one machine and then set the same on other machines. The main reason to define and share these settings is so all common information is consistently defined across all users. The Copy Settings Wizard makes it easy for you to do this.

The sample shown in this chapter is just one way you can do this. In your organization, you may need to save the settings without the keyboard shortcuts. Or perhaps the keyboard shortcuts are the feature you want to transfer to the other computers. Remember, while consistency among your users is important for quality and standards, some individuality can be a good thing to increase productivity.

You can easily change and restore SolidWorks settings for specific tasks, such as switching from your standard settings to settings that specifically apply to consumer product design, machine design, or mold design. Maybe you model some parts in mm and others in inches. Or you create a lot of machined parts then every so often have to model a sheet metal part.

Save Your Interface and Settings

Many people work on multiple computers. Others have multiple users on the same computer. How do you move to another computer and load your interface easily? Even if you only work on one computer, I recommend that after you customize SolidWorks to the way that you like it, you do this to save your settings. You never know when you may inadvertently modify your toolbars or keyboard shortcuts. And this is the fastest easiest way to reset them.

First, close SolidWorks by pulling down the "File" menu and picking **Exit**.

Click on the Windows **Start** button and pick **All Programs – SolidWorks 2007 – SolidWorks Tools – Copy Settings Wizard**.

In the **SolidWorks Copy Settings Wizard** dialog box, click the **Save Settings** button and click **Next**.

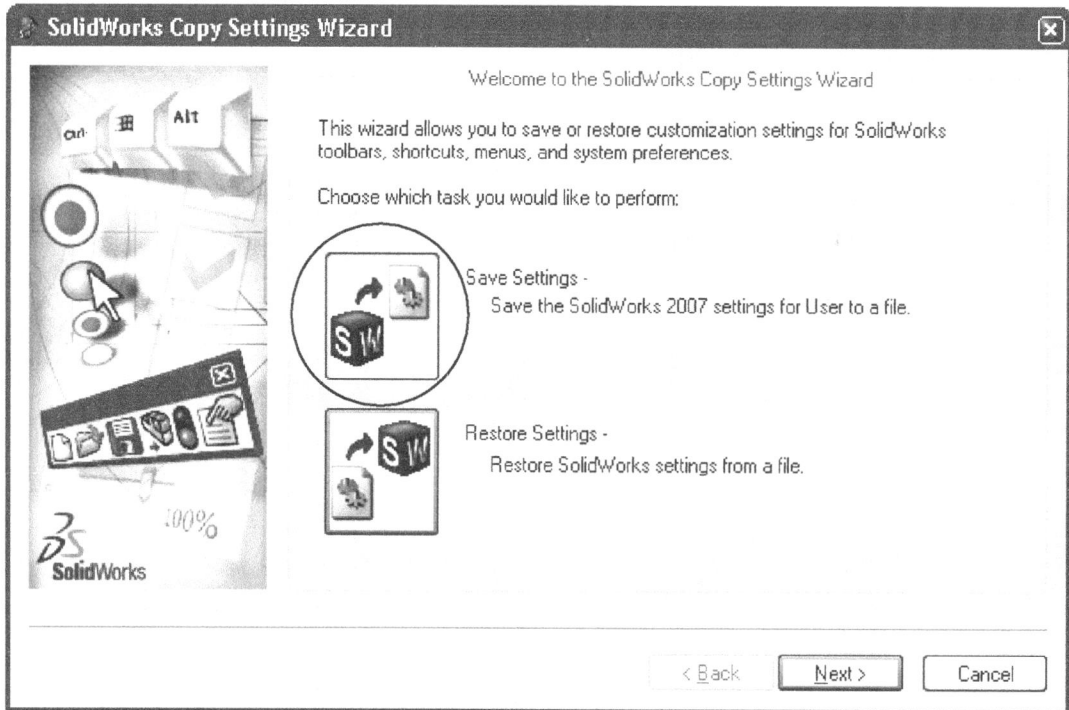

In the **SolidWorks Copy Settings Wizard** dialog box, click the Browse button and browse to your new **SolidWorks Files** folder and enter '**My Settings**' for the name of the settings file and click **Save**. If you have multiple users, change the 'My' to your name, as in '**Neil Settings**'.

Under **Select which settings to save to file**, make sure that all four check boxes are checked and the **All toolbars** radio button is selected. This will save all of the setting.

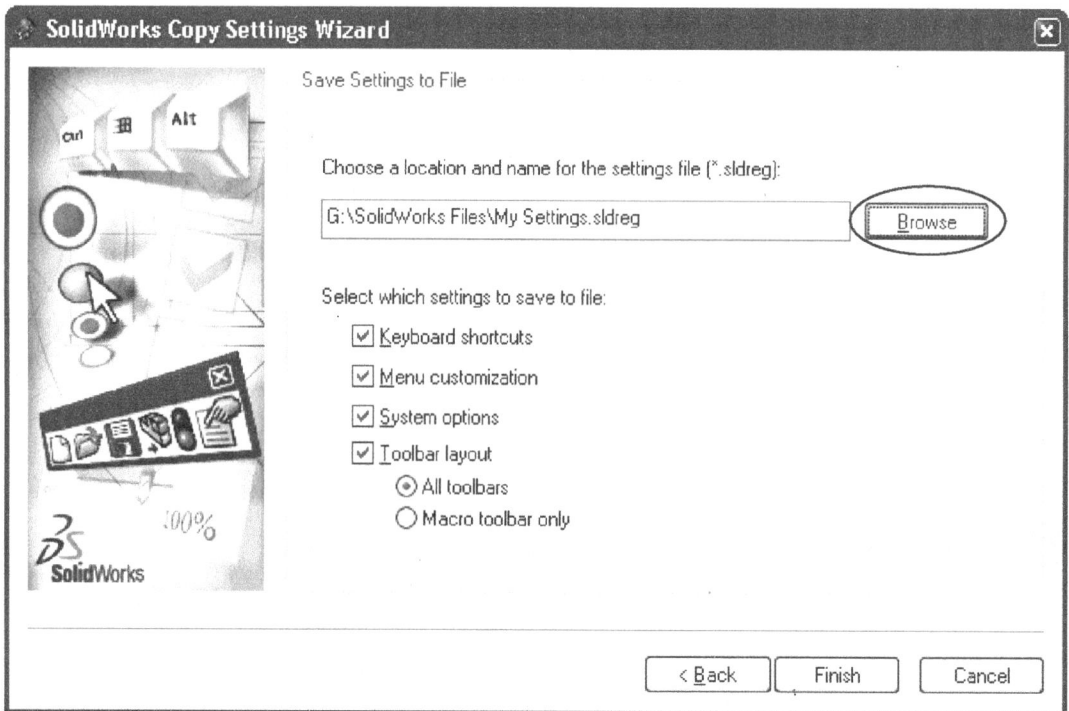

Click **Finish**.

A dialog box will appear, confirming that the settings have been copied successfully to the **My Settings.sldreg** file, and it is saved to the **SolidWorks Files** folder.

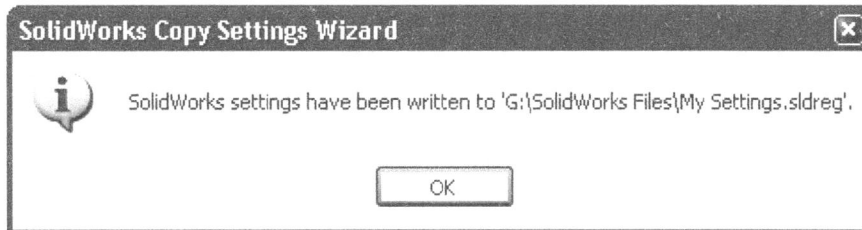

Click **OK**.

Task Pane and Industry Customization

Now that your settings are saved, it does not matter what changes you make to the settings and interface because you can easily restore them to the saved settings. Let's make some changes. SolidWorks has customized the interface for certain areas of expertise, also including changes to the links on the **SolidWorks Resources** tab in the Task Pane and the display of menu items on the "Insert" menu and the "Tools" menu.

To access the built-in customizations, open SolidWorks.

Press **Ctrl+N** or click the **New** button in the "Standard" toolbar.

Double click on **Cedar**.

Right click on the CommandManager and pick **Use Large Buttons with Text**.

Press **Ctrl+Shift+C** or pull down the "Tools" menu and pick **Customize**.

In the **Customize** dialog box, click on the last tab, labeled **Options**.

Under **Work flow customization**, there are three areas of expertise to choose from.

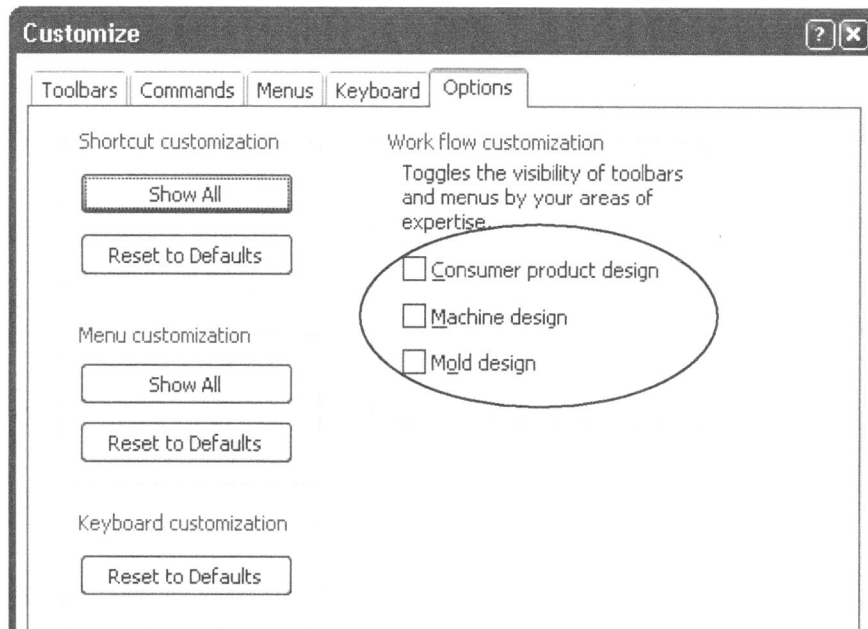

Check **Consumer product design**. The "Surfaces" toolbar is added to the CommandManager.

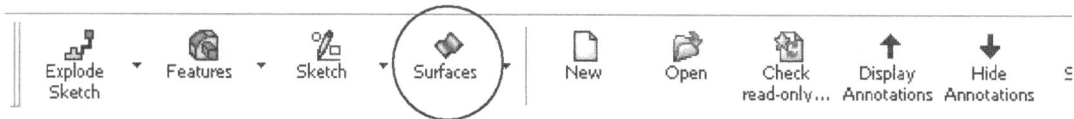

Check **Machine design**. The "Sheet Metal" and "Weldments" toolbars are added to the CommandManager.

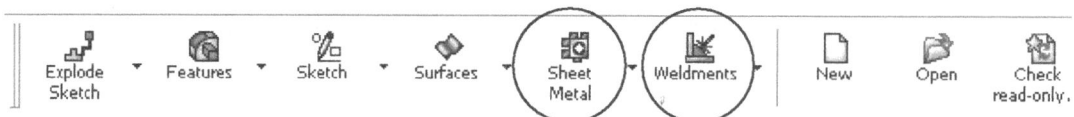

Check **Mold design**. The "Surfaces" and "Molds" toolbars are added to the CommandManager.

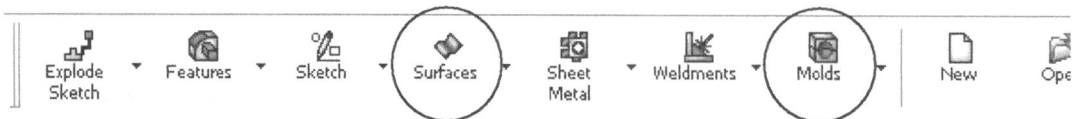

In the **Customize** dialog box, click **OK**.

Restore Your Settings on Any Computer

Now that your settings are all changed, let's change them back to your saved settings. This can be done at any time on any computer. This technique is great if you want to use someone else's computer for a short time and then return it back to the way they had it.

Close SolidWorks by pulling down the "File" menu and picking **Exit**.

Click on the Windows **Start** button and pick **All Programs – SolidWorks 2007 – SolidWorks Tools – Copy Settings Wizard**.

In the **SolidWorks Copy Settings Wizard** dialog box, click the **Restore Settings** button and click **Next**.

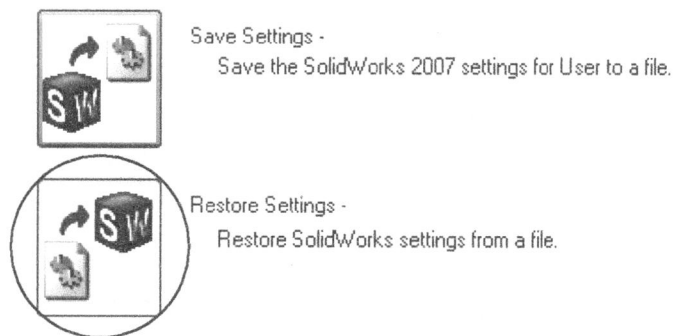

Save Settings -
   Save the SolidWorks 2007 settings for User to a file.

Restore Settings -
   Restore SolidWorks settings from a file.

In the **SolidWorks Copy Settings Wizard** dialog box, browse to the settings file that you want to use to restore the settings. Use **My Settings.sldreg** in the **SolidWorks Files** folder.

Under **Select which settings to restore from the file**, make sure that all four check boxes are checked. (If you just wanted to use your keyboard shortcuts and change nothing else, you would only check **Keyboard shortcuts**).

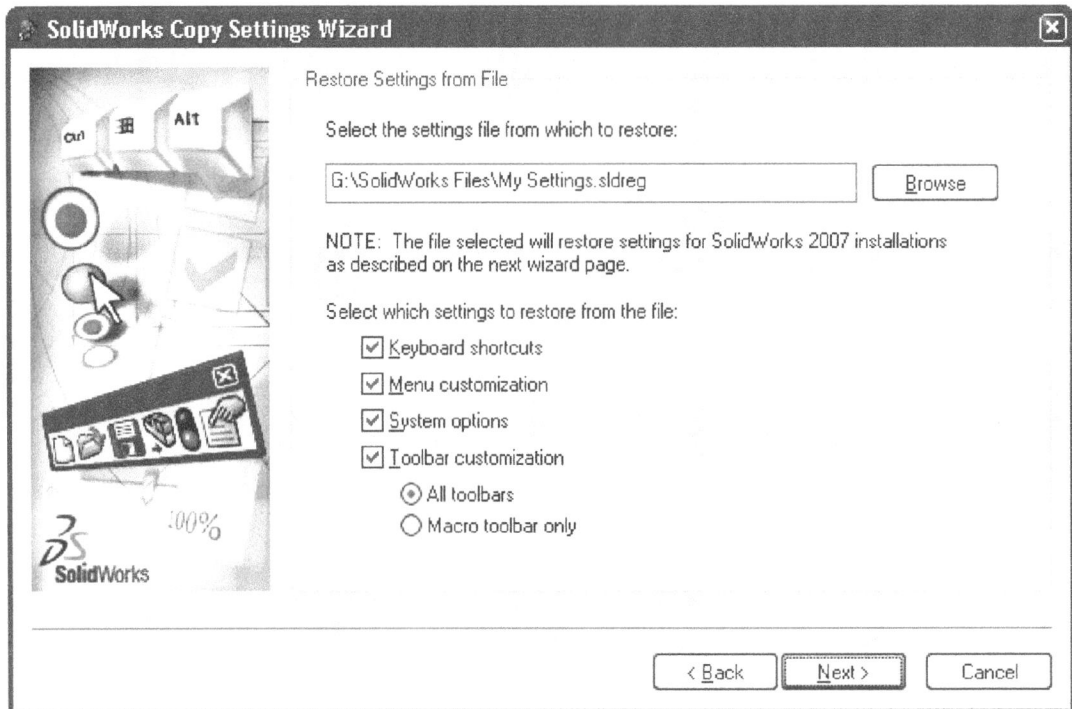

Click **Next**.

In the **SolidWorks Copy Settings Wizard** dialog box, under **Select the Destination**, click on the **Current user** button.

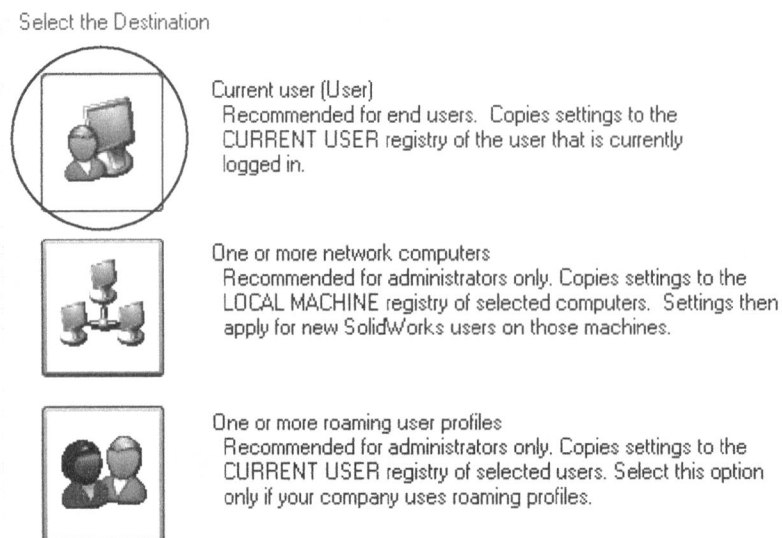

Click **Next**.

In the **SolidWorks Copy Settings Wizard** dialog box, make sure that **Create backup of current settings for User** is checked. With this checked, the current settings are saved as a .sldreg file in the folder chosen earlier. Then, the settings from the selected .sldreg file are restored. This is how you would put your settings on someone else's computer. When you are done, all you have to do is run the Copy Settings Wizard when you are done and pick the backup file to restore. All the settings will be changed back like you were never there.

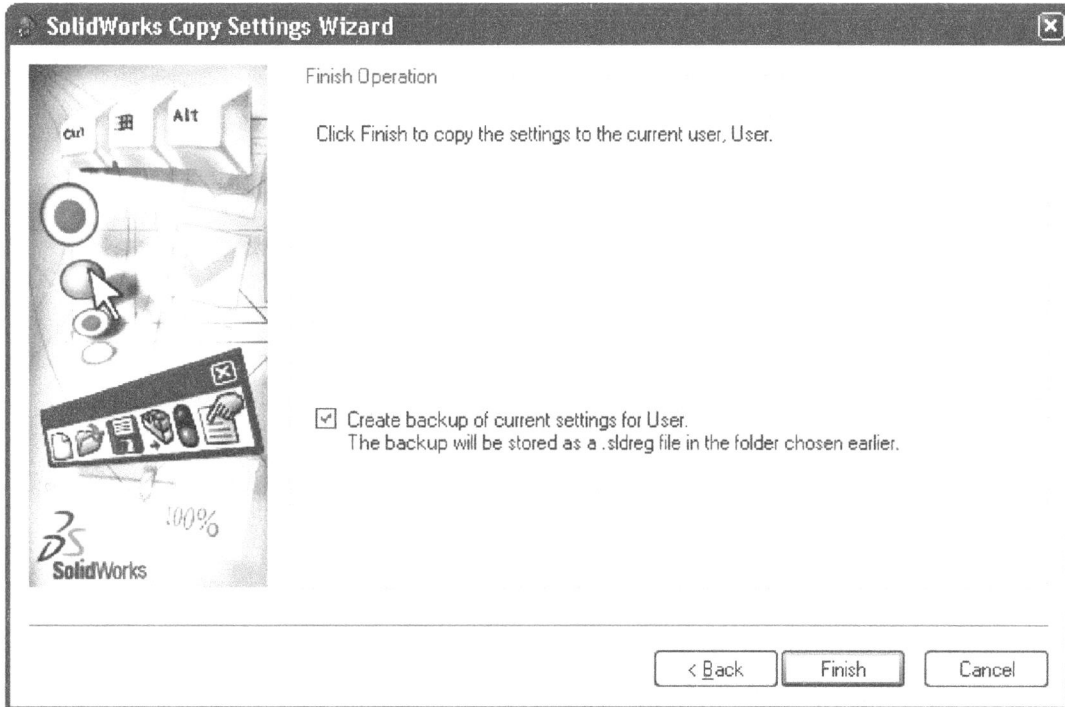

Click **Finish**.

A dialog box will appear confirming that the settings have been copied successfully.

Click **OK**.

Note that you can also update the settings on another computer. In the **SolidWorks Copy Settings Wizard**, pick **One or more network computers** instead of **Current user**. A list of available computers is displayed to choose from.

Pick a user or a computer and click **Add**. Pick as many users or computers as you want, then click **Next**.

Open SolidWorks.

Press **Ctrl+N** or click the **New** button in the "Standard" toolbar.

Double click on **Cedar**.

Note that all your settings have been restored.

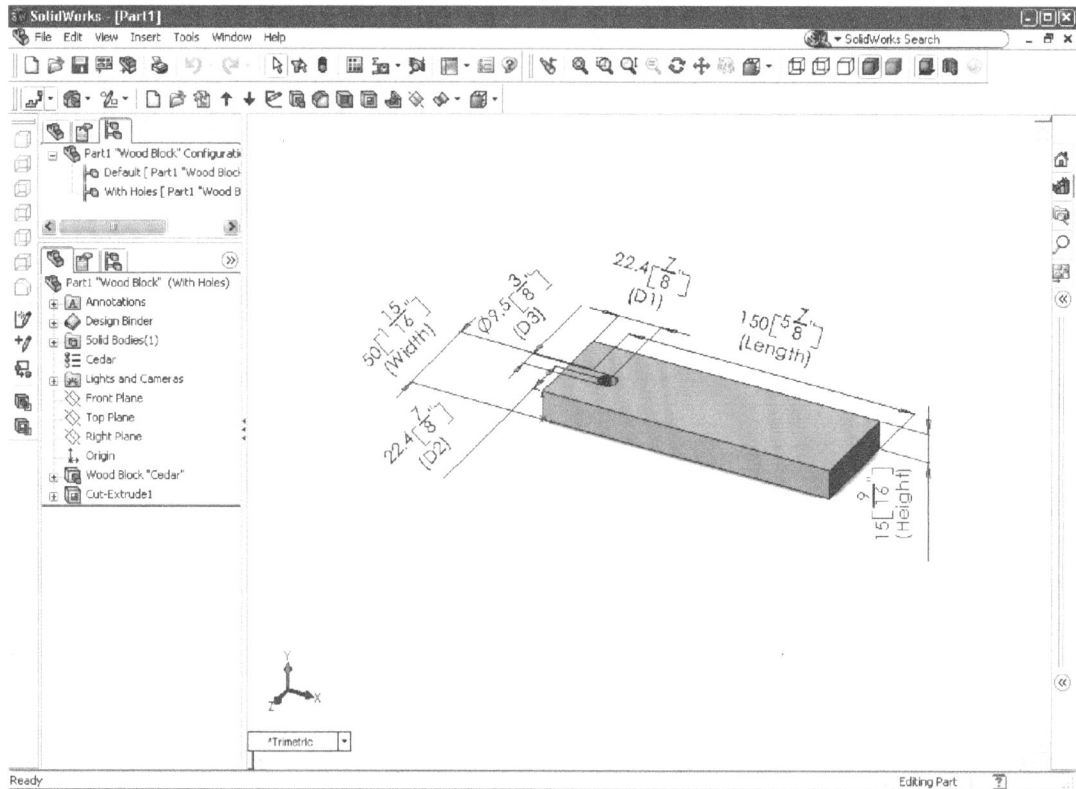

Congratulations

This should get you well on your way to get SolidWorks set up exactly how you want it and be more efficient using the software. If you like to play around with things, tweak your settings, and access the hidden power inside SolidWorks. Then, take what you've learned about customizing features like menus, toolbars, keyboard shortcuts, and views and apply the techniques to your everyday use of SolidWorks. Change the settings and user interface according to your preferences, have fun and become more productive, getting the most out of your SolidWorks.

Let us know what you did or didn't like about this book by emailing us at books@SheetMetalGuy.com. Please be sure to include the book's title as well as your name and contact information.